Immunological and Molecular Diagnosis of Infectious Disease

Immunological and Molecular Diagnosis of Infectious Disease

edited by

Daniel Amsterdam
Roger K. Cunningham
Carel J. van Oss
State University of New York at Buffalo
Buffalo, New York

MARCEL DEKKER, INC. NEW YORK • BASEL • HONG KONG

Library of Congress Cataloging-in-Publication Data

Immunological and molecular diagnosis of infectious disease / edited by Daniel Amsterdam, Roger K. Cunningham, Carel J. van Oss.
p. cm.
"Contributions to the XIIIth International Convocation on Immunology ... held in Buffalo, June 1–5, 1996"—Pref.
Includes bibliographical references and index.
ISBN 0-8247-0092-9 (alk. paper)
1. Communicable diseases—Immunodiagnosis—Congresses.
2. Communicable diseases—Molecular diagnosis—Congresses.
I. Amsterdam, Daniel. II. Cunningham, Roger K. III. van Oss, Carel J.
IV. International Convocation on Immunology (13th : 1996 : Buffalo, N.Y.)
[DNLM: 1. Communicable Diseases—diagnosis—Congresses.
2. Immunologic Tests—Congresses. 3. Molecular Biology—methods--Congresses. WC 100 I3325 1997]
RC113.3.I46 1997
616.9'04756—dc21
DNLM/DLC
for Library of Congress

97-5015
CIP

Originally published as *Immunological Investigations*, Volume 26, Numbers 1&2 (Marcel Dekker, Inc.)

The publisher offers discounts on this book when ordered in bulk quantities. For more information, write to Special Sales/Professional Marketing at the address below.

This book is printed on acid-free paper.

MARCEL DEKKER, INC.
270 Madison Avenue, New York, New York 10016

Current printing (last digit):
10 9 8 7 6 5 4 3 2 1

PRINTED IN THE UNITED STATES OF AMERICA

Preface

This volume comprises the majority of the scientific contributions to the XIIIth International Convocation on Immunology, *Immunological and Molecular Diagnosis of Infectious Disease*, sponsored by The Ernest Witebsky Center for Immunology, State University of New York at Buffalo, held in Buffalo, June 1–5, 1996.

In recent years, new or previously unappreciated problems in the field of infectious diseases have emerged. In addition, new assay methods, particularly those involving DNA and RNA amplification and detection, have been developed for the rapid diagnosis of such diseases. While some of the infectious agents, such as *Helicobacter*, *Borrelia*, hantavirus and prions, have only recently come to prominence, others are previously encountered agents that are causing renewed concern after years of apparent control, such as *Mycobacterium tuberculosis* and hepatitis viruses.

Part I of these Proceedings treats methodologies of Immunological and Molecular Approaches, including DNA and RNA amplification methods, ELISA approaches, rapid diagnosis methods, biosensors, and flow cytometry. Part II deals with the more clinical aspects of Bacterial and Parasitic Infections comprising *M. tuberculosis*, *Borrelia burgdorferi*, *Streptococcus pyogenes*, Chlamydia, and schistosomes. Part III treats Viral Infections, e.g., hantaviruses, hepatitis C virus, cytomegalovirus, herpes viruses, and human immunodeficiency virus. Part IV groups Syndromes of Various Infectious Origins, including prions, chronic fatigue syndrome, and superantigens. All major topics discussed at the XIIIth Convocation are included in this volume, with one notable exception.

That exception is the presentation by Dr. Yi-Fu Zhou, of the Molecular and Cell Biology Laboratory, Cardiology Branch, NHLBI, NIH, Bethesda, Maryland, on "The Potential Role of Cytomegalovirus in Atherosclerosis and in Coronary Restenosis Following Angioplasty." To avoid duplication of publication, the text of Dr. Zhou's contribution could not be incorporated in this volume: the essence of the work by Dr. Zhou et al. was published shortly after his presentation in the New England Journal of Medicine (1). In the late 1980s and early 1990s epidemiological studies initiated in different medical schools in the United States (2) and Europe (2,3) indicated that a significantly high percentage of atherosclerosis and heart disease patients showed prior infection with cytomegalovirus (CMV), as contrasted with normal controls. In 1994, Speir et al. (also from the Cardiology Branch, NHLBI, NIH) proved the existence of a molecular link between CMV and the tumor suppressor protein p53 in coronary restenosis lesions (4). The 1996 paper (1) further strengthens the connection between CMV infection and arterial and heart disease.

While it is not feasible to treat all aspects of modern Immunological and Molecular Diagnosis of Infectious Disease in one volume, the 25 chapters of this work aim to provide highlights of the most recent developments in the clinical diagnosis of several major emerging and re-emerging infectious diseases. This volume points to the increasing reliance on these new technologies for what will soon become routine diagnostic procedures.

Daniel Amsterdam
Roger K. Cunningham
Carel J. van Oss

REFERENCES

1. Y.F. Zhou, M.B. Leon, M.A. Waclawiw, J.J. Popma, Z.X. Yu, T. Finkel, and S.E. Epstein. N. Eng. J. Med. *335*, 624–630 (1996).
2. J.L. Melnick, E. Adam, and M.E. DeBakey. JAMA *263*, 2204–2207 (1990).
3. M. Musiani, M.L. Zerbini, A. Muscari, G.M. Puddu, G. Gentilomi, D. Gibellini, G. Gallinella, P. Puddu, and M. La Placa. Microbiologica *13*, 35–41 (1990).
4. E. Speir, R. Modali, E.S. Huang, M.B. Leon, F. Shawl, T. Finkel, and S.E. Epstein. Science *265*, 391–394 (1994).

ADDENDUM

Due to unforeseen circumstances the contribution by J.B. Myers and D. Amsterdam on "The Laboratory Diagnosis of Cytomegalovirus Infections" does not appear in this volume. However, it has been printed in Immunological Investigations, *26*(*3*), 383 (1997), Marcel Dekker, Inc.

Contents

Contributors

Danièle Altschuh Immunochemistry Department, CNRS, Institut de Biologie Moléculaire et Cellulaire, Strasbourg, France

Daniel Amsterdam Departments of Medicine and Pathology, State University of New York at Buffalo, Buffalo, New York

Donald Armstrong Clinical Laboratory Science, State University of New York at Buffalo, Buffalo, New York

E. Bisaccia Department of Dermatology, Morristown Memorial Hospital, Morristown, New Jersey

J. E. Butler Department of Microbiology, University of Iowa, Iowa City, Iowa

R. I. Carp NYS Institute for Basic Research in Developmental Disabilities, Staten Island, New York

Jean Chatellier Immunochemistry Department, CNRS, Institut de Biologie Moléculaire et Cellulaire, Strasbourg, France

P. K. Coyle Department of Neurology, School of Medicine, State University of New York at Stony Brook, Stony Brook, New York

Yadira Edwards-Spring Lederle-Praxis Biologicals, West Henrietta, New York

Mary Jo Evans Department of Pediatrics, State University of New York at Buffalo, Buffalo, New York

Robert E. Farrell, Jr. Exon-Intron, Inc., Columbia, Maryland

R. Fersko NYS Institute for Basic Research in Developmental Disabilities, Staten Island, New York

Betty A. Forbes Department of Pathology, State University of New York Health Science Center, Syracuse, New York

K. Fujihara Department of Neurology, Roswell Park Cancer Institute and Department of Neurology, State University of New York at Buffalo, Buffalo, New York

Christine Ginocchio Department of Infectious Diseases, North Shore University Hospital, Manhasset, New York

N. Glenister Department of Neurology, Roswell Park Cancer Institute, Buffalo, New York

B. Goldman Department of Dermatology, Roswell Park Cancer Institute and Department of Dermatology, State University of New York at Buffalo, Buffalo, New York

Steven J. Greenberg Department of Neurology, Roswell Park Cancer Institute, and Department of Neurology, State University of New York at Buffalo, Buffalo, New York

M. R. Hammerschlag Departments of Pediatrics and Medicine, State University of New York Health Science Center at Brooklyn, Brooklyn, New York

John Hay Department of Microbiology, State University of New York at Buffalo, Buffalo, New York

D. N. Howell Department of Pathology, Duke University Medical Center and Department of Pathology, VA Medical Center, Durham, North Carolina

E. S. Jaffe Laboratory of Pathology, National Cancer Institute, National Institutes of Health, Bethesda, Maryland

Mark Kaplan Department of Infectious Diseases, North Shore University Hospital, Manhasset, New York

R. J. Kascsak NYS Institute for Basic Research in Developmental Disabilities, Staten Island, New York

A. C. Mawle Division of Viral and Rickettsial Diseases, Centers for Disease Control and Prevention, Atlanta, Georgia

S. E. Miller Departments of Microbiology and Pathology, Duke University Medical Center, Durham, North Carolina

Brian M. Murray Department of Medicine, State University of New York at Buffalo, Buffalo, New York

Jean Myers Virology Laboratory, Erie County Medical Center, Buffalo, New York

P. Navarro Department of Microbiology, University of Iowa, Iowa City, Iowa

Frederick S. Nolte Department of Pathology and Laboratory Medicine, Emory University School of Medicine, Atlanta, Georgia

A. R. Oseroff Department of Dermatology, Roswell Park Cancer Institute and Department of Dermatology, State University of New York at Buffalo, Buffalo, New York

S. Pincus Department of Dermatology, State University of New York at Buffalo, Buffalo, New York

Deborah Povinelli Division of Infectious Diseases, Children's Hospital of Buffalo, Buffalo, New York

D. Pulgiano NYS Institute for Basic Research in Developmental Disabilities, Staten Island, New York

Nathalie Rauffer-Bruyère Immunochemistry Department, CNRS, Institut de Biologie Moléculaire et Cellulaire, Strasbourg, France

Pascale Richalet-Sécordel Immunochemistry Department, CNRS, Institut de Biologie Moléculaire et Cellulaire, Strasbourg, France

Marie Riepenhoff-Talty Department of Pediatrics, State University of New York at Buffalo, Buffalo, New York

Kate Rittenhouse-Diakun Clinical Laboratory Science, State University of New York at Buffalo, Buffalo, New York

Joseph W. Romano Advanced BioScience Laboratories, Inc., Kensington, Maryland

R. Rubenstein NYS Institute for Basic Research in Developmental Disabilities, Staten Island, New York

Hélène Saunal Immunochemistry Department, CNRS, Institut de Biologie Moléculaire et Cellulaire, Strasbourg, France

J. Schachter Department of Laboratory Medicine, University of California, San Francisco, San Francisco, California

P. M. Schlievert Department of Microbiology, Medical School, University of Minnesota, Minneapolis, Minnesota

T. Shanahan Department of Microbiology, State University of New York at Buffalo and Histocompatibility and Immunogenetics Division, Erie County Medical Center, Buffalo, New York

Roxanne N. Shurtliff Advanced BioScience Laboratories, Inc., Kensington, Maryland

Dennis L. Stevens Division of Infectious Diseases, Veterans Affairs Medical Center, Boise, Idaho; and Department of Medicine, University of Washington School of Medicine, Seattle, Washington

Carleton C. Stewart Laboratory of Flow Cytometry, Roswell Park Cancer Institute, Buffalo, New York

J. Sun Department of Microbiology, University of Iowa, Iowa City, Iowa

Victor C. W. Tsang Immunology Branch, Parasitic Diseases Division, National Center for Infectious Diseases, Centers for Disease Control and Prevention, U.S. Public Health Service, Atlanta, Georgia

Marc H. V. Van Regenmortel Immunochemistry Department, CNRS, Institut de Biologie Moléculaire et Cellulaire, Strasbourg, France

Alice Wendt Division of Infectious Diseases, Children's Hospital of Buffalo, Buffalo, New York

T. Ulf Westblom Department of Internal Medicine, Texas A&M University, College of Medicine and Department of Veterans Affairs, Central Texas Veterans Health Care System, Temple, Texas

Judith C. Wilber Nucleic Acid Diagnostics and New Markers, Chiron Corporation, Emeryville, California

Patricia P. Wilkins Immunology Branch, Parasitic Diseases Division, National Center for Infectious Diseases, Centers for Disease Control and Prevention, U.S. Public Health Service, Atlanta, Georgia

Kimberly G. Williams Advanced BioScience Laboratories, Inc., Kensington, Maryland

Joseph J. Zambon School of Dental Medicine, State University of New York at Buffalo, Buffalo, New York

Xinyang Zhao Department of Microbiology, State University of New York at Buffalo, Buffalo, New York

PART I

IMMUNOLOGICAL AND MOLECULAR APPROACHES

DNA AMPLIFICATION

Robert E. Farrell, Jr.
Exon-Intron, Inc.
9151 Rumsey Road, Suite 130
Columbia, Maryland 21045

ABSTRACT

The polymerase chain reaction has become a mainstream tool for the molecular biologist. The sensitivity, efficiency, and speed of this method is unparalleled for the amplification and detection of exquisitely minute quantities of nucleic acids. Through repetitive cycles of heat denaturation of samples, followed by the base pairing of primers designed to identify one DNA sequence among the cellular heterogeneity, and finally synthesis of new DNA strands identical to the target, single molecules and individual genes can be detected and subsequently characterized. This method has revolutionized the study of gene organization, structure, and expression, not to mention offering newer, faster, and more economical means for the clinical detection infectious disease.

That PCR has been fruitful is undisputed; however, the method is not without shortcomings. Among the major limitations of this method are the absolute requirement for well-designed primers, the super sensitivity of this method to biological contaminants from any of a variety of sources, and subtle, though very important, inter- and intra-laboratory variations in technique.

UNDERSTANDING PCR

The Polymerase Chain Reaction[1] (PCR) has, without a doubt, revolutionized the entirety of molecular biology. Nearly every research project, application, and diagnostic methodology has implemented molecular cloning and identification at some level, and PCR has become, perhaps, the premier tool.

Recent refinements in PCR-related techniques have afforded an unprecedented level of sensitivity and resolution in most molecular assays. As a direct result of the sensitivity and resolving power of this technique, often only a few short hours span the beginning of the search for a new gene and the selection of the gene itself. The technique of PCR has obvious advantages, and the simple methodology has been adapted in order to develop assays for the identification of single copy gene sequences from complex genomes, the selection of

[1]The Polymerase Chain Reaction process is covered by patents issued to Hoffman-LaRoche, Inc. Use of the PCR process requires a license under their patents.

individual members of multigene families, identification and deliberate introduction of point mutations, deletions, amplifications, and gene rearrangements, and the identification of infectious agents by amplification of foreign genetic material from the blood of the host.

Comprehensive reviews of the subtleties of PCR have been published elsewhere (1,2,3), and a brief review of the method is presented here in order to convey the profound impact PCR continues to have on the daily routine of the most sophisticated and not-so-sophisticated laboratories around the world. The method of PCR is primer-mediated enzymatic amplification of specific genomic or cDNA sequences. Succinctly, PCR is possible because thermostable prokaryotic DNA polymerases are readily available, because single stranded DNA molecules (oligonucleotides) can be economically synthesized consisting of any combination of nucleotides desired by the investigator, because of instrumentation capable of efficient regulation of temperature, and because of DNA sequencing technology.

The polymerase chain reaction is divided into cycles, each of which is comprised of three discrete components, namely **denaturation** of both target sequences and primers, by which all of these molecules become single-stranded; **primer annealing**, by which the oligonucleotide primers involved base pair to a complementary sequence within the template (target) DNA; and **primer extension**, in which the polymerase activity of a thermostable enzyme (e.g. *Taq* polymerase) synthesizes DNA from the 3′-OH terminus contributed by each primer molecule which has base-paired to the template.

PCR: THE CYCLE

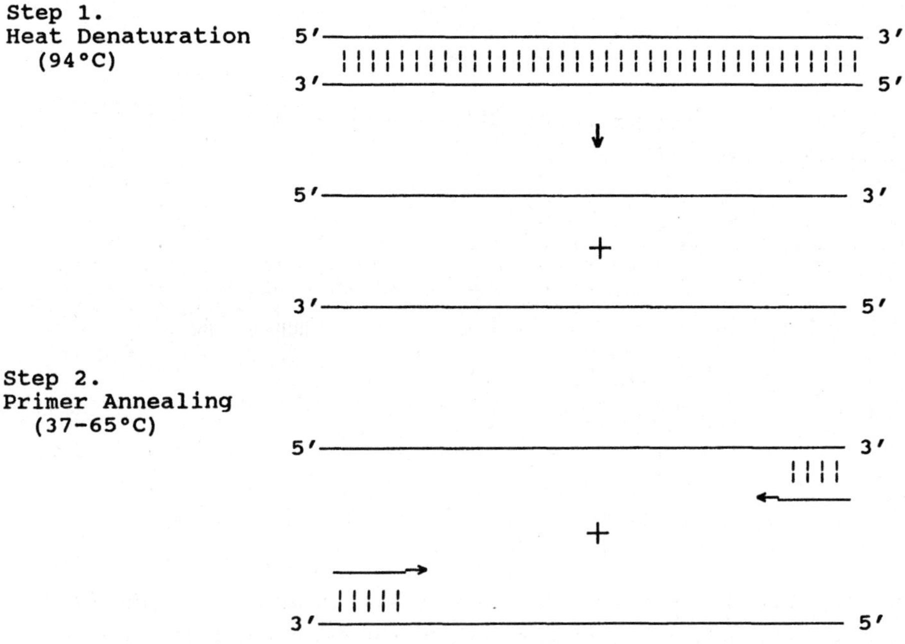

Step 3.
Primer Extension
(72°C)

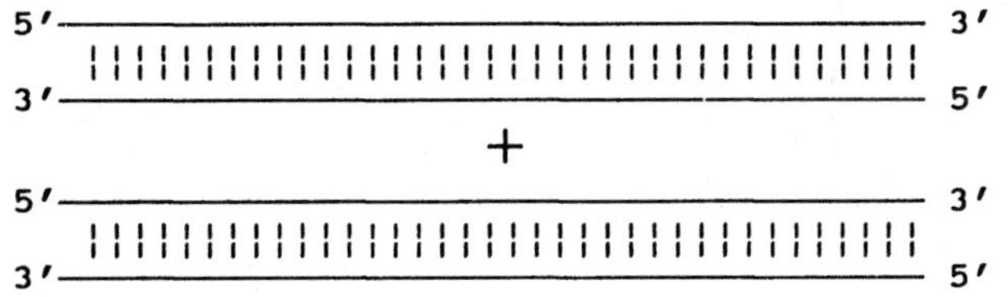

These three steps, as illustrated above, constitute "cycle 1" and the repetition of these steps using the products of the prior cycle as new template supports the geometric accumulation of those sequences. After only 30 cycles it is thus possible to achieve a multimillion fold amplification of template (2^{30} = 1,073,741,824; however the reaction is not 100% efficient).

The ability to perform PCR requires some knowledge of the sequence to be amplified or of the sequences which flank the region to be amplified. In the absence of direct nucleotide sequence information, the amino acid sequence of the corresponding peptide can be scrutinized in order to deduce the probable nucleotide sequence responsible for encoding that particular combination of amino acids. If this strategy is employed for primer design, a lack in the completeness of protein sequence data may not be an impediment because as few as 6 or 7 sequenced amino acids may be adequate for the design of an 18- or 21-mer oligonucleotide primer. Quite useful in this regard are protein domains containing large amounts or methionine and tryptophan, both of which are single codon amino acids. If such a domain has not been identified, eight additional amino acids have only two possible codons, a density of which may facilitate primer design. Thus, it may be possible to deduce an appropriate primer sequence to support PCR. Obviously, this is not a fool-proof method of cloning by PCR, however, because of the fair amount of degeneracy within the code itself (some amino acids have more than one possible codon).

Fortunately, oligonucleotide primers are not required to anneal or match perfectly with the template. Were this an absolute requirement, the PCR process would be far less productive and far more frustrating indeed. Because of the extremely short nature of the oligonucleotide primers, the thermodynamic behavior of these molecules and their affinity for each other can be predicted with great accuracy. Often, temperature deviations of as little as one or two degrees mean the difference between successful amplification and the failure of that reaction. By lowering the temperature of the primer annealing component of each cycle (above) in the polymerase chain reaction, primers which are not perfectly base paired (matched) with their intended templates will be capable of base pairing among complementary nucleotides and at the same time will tolerate a mismatch at those locations where non-complementary nucleotides abut. The extent of base pairing is less critical than the location of mismatches; base pairing of the terminal 3′ dinucleotide is the only absolute requirement to support amplification.

QUANTIFICATION

Among the more popular applications of PCR technology is the method of transcript amplification for both sequencing and quantification purposes. Known as RNA PCR or RT-PCR, this method involves reverse transcriptase-mediated cDNA synthesis according to standard methodologies, using RNA purified according any of a variety of protocols (4), followed by PCR amplification of the newly synthesized cDNA. The mass of a PCR product so generated is directly proportional to the relative abundance of the transcript.

LIMITATIONS

While the promise of PCR has been fruitful indeed, the method does have limitations. A great many questions could certainly be answered, were there primers available to support analysis by PCR. A fundamental difficulty in many laboratories is the inability to perform PCR because the required primers have yet to be developed. This is especially true of research which involves the study of a "biological activity" of a yet-to-be characterized enzyme or other gene product. As stated above, the ability to perform PCR is dependent upon having some nucleic acid sequence information either about the sequence which is to be amplified, or about the sequences which flank the sequence to be amplified.

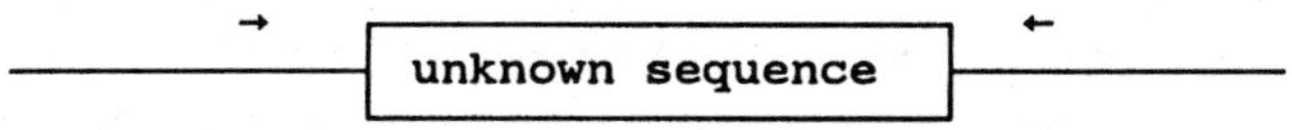

This sequence information can be either direct or indirect: direct sequences are derived by sequencing at least part of the target to be amplified, while indirect sequence information may be derived from another member of a multigene family, by characterization of the same gene in a different species (cloning by evolutionary relatedness), or by direct protein sequencing. The bottom line: no primers, no PCR.

That the polymerase chain reaction is too sensitive is a frequent complaint among investigators who routinely perform PCR. Given that a geometric amplification of template occurs, a lone carry-over molecule is enough to generate false signals and/or false positives. The gravity of such an occurrence is especially profound in a diagnostic setting for a number of obvious reasons. Contamination in the PCR environment has been traced to aerosol formation from micropipettors, tainted stock solutions of PCR reagents, contaminated thermal cyclers, lab coats and, in some cases, the epidermal layers of the investigator himself.

Lastly, the PCR technique requires empirical formulation of the exact conditions for amplification with each primer pair brought into the laboratory. Magnesium, a required divalent cation, is notorious for mediating the mispriming and amplification of sequences unrelated to the template and, by virtue of its ability to quantitatively bind dNTP's, exert great influence on the efficiency of amplification in general. Moreover, the precise type of thermal cycler that is utilized may have a profound bearing on the efficiency and outcome of the reaction. Because of the way in which samples are heated and cooled, and the ramping time when moving among temperatures, enormous variation should be expected when comparing similar/identical protocols performed in different laboratory settings.

PCR has, without a doubt, revolutionized molecular biology, and all related disciplines. Experiments which in the past required days/weeks to perform can now be performed in a matter of hours. The ability to confirm a diagnosis by sensitive, specific means of DNA amplification affords the physician an opportunity to intervene more rapidly upon suspicion or clinical manifestation of infection or a disease state.

REFERENCES

1. Innis, M.A. et al., PCR Protocols. Academic Press, San Diego, California (1990).

2. McPherson, M.J. et al., PCR: A Practical Approach. Oxford University Press, Cambridge, England (1995).

3. Clewley, J.P. The Polymerase Chain Reaction (PCR) for Human Viral Diagnosis. CRC Press, Boca Raton, Florida (1994).

4. Farrell, R.E. RNA Methodologies: A Laboratory Guide for Isolation and Characterization. Academic Press. San Diego, California (1993)

BRANCHED DNA FOR QUANTIFICATION OF VIRAL LOAD

Judith C. Wilber
Clinical Research
Nucleic Acid Diagnostics and New Markers
Chiron Corporation
4560 Horton Street
Emeryville, California 94608, USA

ABSTRACT

This is a summary of a presentation made at the 13th International Convocation on Immunology. Nucleic acids in patient samples can be quantified directly using a solid phase nucleic acid hybridization assay based on branched DNA (bDNA) signal amplification technology. For example, HIV RNA is detected in a plasma sample by hybridization of multiple specific synthetic oligonucleotides to the target, 10 of which capture the target onto the surface of a microwell plate and 39 of which mediate hybridization of branched DNA molecules to the *pol* region of each HIV RNA molecule. Alkaline phosphatase-labeled probes bind to each arm of the branched DNA molecules. Detection is achieved by incubating the complex with a chemiluminescent substrate and measuring the light emission. The signal is directly proportional to the level of target nucleic acid, and the quantity of HIV RNA in a sample is determined by comparison with a 4-point standard curve. In order to ensure that different subtypes of HIV-1 were detected and quantified equally, in vitro RNA transcripts of the *pol* region of HIV subtypes A-F were purified and quantified by OD 260, phosphate analysis, and hyperchromicity. These characterized transcripts were then quantified using the bDNA assay. Comparisons were made using a ratio of signal per attomole for each transcript. Genetic subtypes A-F quantified within a factor of 1.5, indicating that the bDNA assay can be used to measure viral load in clinical samples regardless of genotype. Accuracy is important because several studies indicate that there may be a threshold level of virus which predicts progression of HIV disease. Detection of change in viral load is important in determining the efficacy of therapy. The bDNA assay for HCV RNA can be used to determine level of virus in HCV-infected individuals and assist in establishing prognosis prior to initiation of alpha-interferon therapy. Patients with lower levels of virus are more likely to have a sustained response to therapy. Patients who respond to treatment typically have a rapid decline in virus load within one to four weeks of the start of therapy. Many patients relapse when therapy is discontinued as evidenced by a rise in virus load to near pre-treatment levels. Sustained response is most often seen with patients who have lower pre-treatment levels of RNA.

METHOD

The branched DNA (bDNA) assay is a quantitative signal amplification method based on a series of specific hybridization reactions and chemiluminescent detection of hybridized probes in a microwell format. It involves binding of the target nucleic acid to the surface of a microwell plate by a series of probes that are complementary to capture probe on the plate, the other part of the probe being specific for the target. Another series of probes is added, designed to be complementary to the target and also bind synthetic branched DNA molecules. Each bDNA molecule has 15 arms, each arm able to hybridize to 3 alkaline phosphatase molecules. Thus, 45 alkaline phosphatase molecules can bind to each bDNA (Figure 1). In the HCV assay, 18 bDNA molecules are bound to each HCV RNA molecule in the 5' untranslated and core regions, and in the HIV assay 39 bDNA molecules are bound to each HIV RNA molecule in the *pol* region.

In order to ensure that different subtypes of HIV-1 were detected and quantified equally, *in vitro* RNA transcripts of the *pol* region of HIV subtypes A-F were purified and quantified by OD 260, phosphate analysis, and hyperchromicity. These characterized transcripts were then quantified using the bDNA assay. Comparisons were made using a ratio of signal per attomole for each transcript. Genetic subtypes A-F quantified within a factor of 1.5, indicating that the bDNA assay can be used to measure viral load in clinical samples regardless of genotype. As a result of a similar study of the HCV bDNA assay, the target probes were redesigned to ensure that each of the HCV genotypes is detected equally (1). The redesign increased the overall clinical sensitivity to over 95 percent (2). There was an overall percent coefficient of variation of 21-24% when over five hundred replicates of three different samples were tested over ten monthswith six lots of reagents by three different operators.

HEPATITIS

Hepatitis C virus (HCV), hepatitis B virus (HBV) can be quantified directly from serum or plasma using a solid phase nucleic acid hybridization assay based on branched DNA (bDNA) signal amplification technology. Assessment of serum alanine aminotransferase (ALT) and other liver enzyme measurements are markers of liver disease but not viral activity. Quantification of viral load has the advantage of providing information on viral kinetics and provides a greater knowledge of the disease process, as it yields a direct measurement of viremia.

HCV RNA baseline data obtained utilizing bDNA analysis of stored samples from interferon-treated patients has provided confirmation of the relationship between initial viral load and patient response to therapy. A sustained response was defined as normalization of serum ALT for at least 6 months after discontinuation of therapy. An analysis was performed on a series of patients from several worldwide studies of standard interferon therapy, where pre-treatment HCV RNA viral load was measured by bDNA. The results demonstrated that there was no significant difference in initial viral load between non-responders and responders who relapsed. The majority of long-term responders, however, had very low baseline viral load. An increasing percentage of sustained response was noted in individuals with lower pre-treatment viral RNA levels. Quantification of HCV RNA by bDNA assay as a predictor of sustained response to administration of interferon in an HCV-infected population illustrates that in the entire population, 19 percent will have a sustained response to therapy, with this figure increasing as high as 49 percent as viral load decreases. Knowing the predicted response to interferon therapy may help set expectations for the patient. It is possible that those with high viral levels may benefit from higher or more frequent doses or longer term therapy.

Fluctuations in viral load are relatively insignificant in untreated patients with chronic HCV infection. Changes of threefold or 0.5 log or greater are rarely seen (3). In a subset of patients, however, RNA and ALT levels can fluctuate considerably, becoming stable only after a number of years (4). Fluctuations are also seen more frequently post-therapy. These observations have led to the theory that improved treatment efficacy may be achieved if treatment or re-treatment is initiated at a time point when a patient's viral load is at its lowest. This possibility is especially important for patients who have already received interferon therapy and have suffered relapse, as experience shows that a large proportion of these patients will relapse again. A small study was conducted in Japan and reported by Koga, Gastroenterology Seminars in 1993. Therapy-naïve individuals were grouped prior to initial treatment by their HCV RNA levels determined by bDNA, i. e. greater than 10 million genome equivalents/ml (n=10), 1 to 10 million (n=15) and less than 1 million (n=11). All participants responded to therapy but eventually relapsed. Viral levels were noted at the time of re-treatment, and, regardless of the individuals' pre-therapy levels, the only patients who experienced a sustained response were those who had viral loads of less than 1 million genome equivalents/ml at the start of re-treatment. Studies are now directed toward use of viral load in making decisions regarding the timing of initiation of treatment or re-treatment in hepatitis C-infected individuals.

At the end of therapy, HCV RNA is often undetectable in the serum, even though many of these patients relapse. Determining whether a patient has truly cleared virus can be problematic due to residual viral replication in the liver. If a follow-up liver biopsy is performed to determine improvement in histology, examination of the liver biopsy material for HCV RNA could be a way of establishing a true patient response. A modification of the bDNA assay can be used to quantify HCV RNA in liver tissue, requiring only simple RNA extraction from liver biopsy samples, followed by the standard microplate assay (5). Since pure nucleic acid is not required for the bDNA procedure, the extraction procedure was optimized for RNA recovery, yielding results that reflect the concentration of the virus in the liver of the patient. In contrast, reverse transcription-PCR (RT-PCR) has been found to be inhibited by heme, present in high concentration in liver tissue, which necessitates purification of RNA and results in loss of quantitative recovery.

HIV

Using bDNA, individuals classified as long-term survivors have been shown to have lower a viral load than those who progress more rapidly (6). In a study using the bDNA assay to measure HIV RNA in plasma of individuals enrolled in MACS, a large cohort study, it was shown that those with high levels of HIV RNA progressed to AIDS more rapidly than those who maintained low levels of RNA (7).

Many HIV-infected individuals have high CD4 levels and high viral load (8). It may be useful to stratify patients by viral load rather than CD4 level for entry into clinical trials. The bDNA assay is currently being used in many laboratories to monitor changes in levels of HIV RNA for several antiviral treatment trials. Two such studies led to new insight into HIV pathogenesis, indicating that HIV replication in vivo is continuous and highly productive (9, 10). It was suggested that because of the rapid turnover of HIV in plasma, protocols for monitoring the antiviral activity of new compounds should be modified to focus on the first few days following drug initiation. New recommendations on the use of HIV RNA quantification include two baseline measurements 2 to 4 weeks apart, a measurement 3 to 4 weeks after initiating a new antiviral therapy, and routine measurements every 3 to 4 months when CD4 counts are done (11). A decrease in HIV RNA of at least 0.5 log is considered incative of antiviral activity.

NEW DEVELOPMENTS

It is often noted that the bDNA assay is not as sensitive as target amplification methods. However, the achievable analytical sensitivity is as high as PCR. The bDNA procedure is similar to an enzyme immunoassay, and the sensitivity is limited by background. In order to reduce non-specific hybridizations, two new bases, iso-cytosine (iso-C) and iso-guanosine (iso-G), were synthesized and utilized in the development of new probes (12). These bases form strong hybridizations with each other, but, as they are not naturally occurring bases, they do not hybridize with natural guanosine and cytosine. In this manner, nonspecific hybridization between the bDNA amplifier and the capture probes has been reduced by substituting some cytosine and guanosine bases with iso-C and iso-G. This method has been used together with modifications in probe design and multi-site preamplification molecules in a bDNA assay for HIV RNA, increasing the sensitivity by more than two orders of magnitude.

REFERENCES

1. M. L. Collins, C. Zayati, J. J. Detmer, B. Daly, J. A. Kolberg, T.-A. Cha, B. D. Irvine, J. Tucker and M. S. Urdea, Analyt. Biochem. 226, 120-129 (1995).

2. J. Detmer, R. Lagier, J. Flynn, C. Zayati, J. Kolberg, M. Collins, M. Urdea and R. Sanchez-Pescador, J. Clin. Microbiol. 34, 901-907 (1996).

3. T. T. Nguyen, A. Sedghi-Vaziri, L. B. Wilkes, T. Mondala, P.J. Pockros, K. L. Lindsay and J. G. McHutchison, J. Viral Hepatitis 3, 75-78 (1996).

4. H. J. Alter, R. Sanchez-Pescador, M. S. Urdea, J. C. Wilber, R. J. Lagier, A. M. Di Bisceglie, J. W. Shih and P. D. Neuwald, J. Viral Hepatitis 2, 121-132 (1995).

5. E. Coehlo-Little, L. J. Jeffers, M. Bartholomew, K. R. Reddy, E. R. Schiff and P. J. Dailey, J. Hepatol. 22, 248 (1995).

6. Y. Cao, L. Qin, L. Zhang, J. Safrit and D. D. Ho, N. Engl. J. Med. 332, 201-208 (1994).

7. J. W. Mellors, C. R. J. Rinaldo, P. Gupta, R. M. White, J. A. Todd and L. A. Kingsley, Science 272, 1167-1170 (1996).

8. C. Pachl, J. A. Todd, D. G. Kern, P. J. Sheridan, S.-J. Fong, M. Stempien, B. Hoo, D. Besemer, T. Yeghiazarian, B. Irvine, J. Kolberg, R. Kokka, P. Neuwald and M. S. Urdea, J. Acquired Imm. Def. Syndromes and Human Retrovirol. 8, 446-454 (1995).

9. D. D. Ho, A. U. Neumann, A. S. Perelson, W. Chen, J. M. Leonard and M. Markowitz, Nature 373, 123-126 (1995).

10. X. Wei, S. K. Ghosh, M. E. Taylor, V. A. Johnson, E. A. Emini, P. Deutsch, J. D. Lifson, S. Bonhoeffer, M. A. Nowak, B. H. Hahn, M. S. Saag and G. M. Shaw, Nature 373, 117-126 (1995).

11. M. S. Saag, M. Holodniy, D. R. Kuritzkes, W. A. O'Brien, R. Coombs, M. E. Poscher, D. M. Jacobsen, G. M. Shaw, D. D. Richman and P. A. Volberding, Nature Med. 2, 625-629 (1996).

12. T. Horn, C-A. Chang and M. L. Collins, Tetrahedron Letters 36, 2033-2036 (1995).

NASBA TECHNOLOGY: ISOTHERMAL RNA AMPLIFICATION IN QUALITATIVE AND QUANTITATIVE DIAGNOSTICS

Joseph W. Romano, Kimberly G. Williams, Roxanne N. Shurtliff
Advanced BioScience Laboratories, Inc.
5510 Nicholson Ln. Kensington, MD 20895, USA

Christine Ginocchio, Mark Kaplan
Department of Infectious Diseases, North Shore University Hospital
300 Community Drive, Manhasset, NY 11030, USA

ABSTRACT

Nucleic acid amplification technologies allow for the development of highly sensitive and specific diagnostic assays. The capacity to amplify and detect analyte targets, which may be present in a clinical sample as a single copy, is characteristic of many of these amplification technologies. NASBA is an isothermal method of nucleic acid amplification with such capability, and is particularly well suited for the amplification of RNA analytes. NASBA utilizes the coordinated activities of three enzymes (AMV-RT, RNase H, T7 RNA polymerase), and two oligonucleotide primers which are specific for the analyte target. The amplification process is part of a total system which includes a versatile nucleic acid isolation procedure, and powerful detection methodology. In this report, the development of NASBA technology for the detection of human Retrovirus RNA will be discussed. Specifically, a qualitative NASBA assay for the RNA of HTLV I, and a quantitative NASBA assay for HIV-1 will be described.

INTRODUCTION

The capacity to detect and quantitate specific RNA analytes is clinically relevant for several reasons. For example, the capacity to detect RNA is critical in determining the efficacy of an antimicrobial therapy. In the case of bacteria, RNA serves as a marker for viable organisms. Since the DNA of an infecting bacteria may persist despite the fact that the organism is no longer viable, DNA is a less useful analyte for this purpose. Therefore, the effectiveness of a therapy can be determined through the post-treatment screening for bacterial RNA. In cases of viral infection, RNA serves as a marker for the intracellular replication of the virus. RNA is also the genetic material of several pathogenic viruses (e.g., Retroviruses, Enteroviruses, etc.). Thus, antiviral therapies can also be evaluated with assays that target viral RNA. Further, the expression of specific cellular transcripts is indicative of particular disease states. This has recently been clearly demonstrated for several forms of cancer (10). It is also possible to conduct genetic screening using RNA targets. An additional benefit of targeting expressed cellular RNA transcripts for genetic or disease diagnoses is the fact that these analytes are typically present in multiple copies, whereas their DNA counterparts are usually present as two copies.

NASBA is an isothermal method of nucleic acid amplification which is most useful in the amplification of RNA analytes since reverse transcription is directly incorporated into the amplification pathway. The process involves three enzymes and two oligonucleotide primers (11). Briefly, the initial event is the annealing of the first primer (P1; antisense) to the RNA analyte. The 3' half of the primer is complimentary to the analyte; the 5' half encodes the T7 RNA polymerase promoter. After primer annealing, avian myeloblastosis virus reverse transcriptase (AMV-RT) extends the primer, producing a cDNA copy of the RNA analyte. The RNA portion of the hybrid is destroyed via the action of RNase H, allowing the second primer (P2; sense) to anneal to the newly synthesized cDNA. The DNA dependent DNA polymerase activity of AMV-RT is now engaged, resulting in the extension of P2, creating a double stranded cDNA copy of the original RNA analyte. Importantly, this cDNA encodes a T7 RNA polymerase promoter at one end. This then serves as a substrate for T7 RNA polymerase, which associates with the promoter and begins to transcribe large amounts of antisense RNA corresponding to the original RNA target. This antisense RNA transcription product can then serve as a template for additional amplification in the process, however the primers will anneal in reverse order (P2 followed by P1). It is this cyclic phase of the process which is responsible for the enormous level of amplification.

Since the entire NASBA amplification process is isothermal, there is no need for temperature cycling. The typical level of amplification is a factor of 10^9. The amplified product is single stranded RNA, and can therefore be readily subjected to different probe hybridization analyses without the need for a denaturing step. Although the NASBA amplification process can be applied to double stranded DNA analytes, two heat denaturing steps are required. In the absence of these heating steps, the process is completely specific for single stranded RNA targets. Thus, background DNA will not be coamplified.

It is the appropriateness of NASBA for RNA which has led to its extensive development as a retroviral diagnostic assay. A quantitative form of the NASBA assay for HIV-1 has been available for some time (14). More recently, we have developed a qualitative assay for HTLV I RNA. Both of these NASBA assays utilize the same nucleic acid isolation procedure (2). However, different detection technologies are used for the different configurations of these assays. In this report, a description of the component technologies comprising the total NASBA system will be presented. The strategy utilized in the quantitative HIV-1 assay, as well as performance data for this assay, will be provided. Finally, the qualitative NASBA assay for HTLV I RNA will be described.

MATERIALS AND METHODS

Nucleic Acid Isolation. The procedure for nucleic acid isolation is the same for both qualitative and quantitative forms of the NASBA assay. The procedure makes use of the guanidine isothiocyanate (GuSCN)-acidified silica method of Boom et al. (2). Briefly, one volume of clinical specimen (whole blood, plasma, serum, sputum, cells, tissue homogenate, CSF, etc.) is added to 9 volumes of lysis buffer (5.25 M GuSCN, 50 mM Tris, pH 7.2, 20 mM EDTA, 1.3% Triton X-100). This inactivates infectious virus and stabilizes the nucleic acids by denaturing nucleases. Next, 50 uL of acid treated silica (1 gm/mL) is added to the lysate. The liberated nucleic acids (single and double stranded forms of RNA and DNA) associate in solid phase with the silica. The silica:nucleic acid complex is pelleted by centrifugation and washed repeatedly (twice with 5.25 M GuSCN, 50 mM Tris, pH 6.4, 20 mM EDTA; twice with 70% ethanol, and once with acetone). After complete volitalization of the acetone from the silica pellet, the nucleic acids are eluted from the silica into water.

Amplification by NASBA. The basic NASBA amplification procedure is the same for both quantitative and qualitative configurations of the assay. The procedure is a modified version of that described by van Gemen et al. (1994). To 5 uL of nucleic acid extract, 10 uL of 80 mM Tris, pH 8.3, 24 mM $MgCl_2$, 140 mM KCl, 10 mM DTT, 2.0 mM each dNTP, 4.0 mM each NTP, 30% DMSO, 0.4 uM P1, 0.4 uM P2) are added. This mix is heated to 65^oC for 5 minutes, then cooled to 41^oC. Once cool, 5 uL of enzyme mix (6.4 units/uL T7 RNA polymerase, 1.3 units/uL AMV-RT, 0.02 units/uL RNase H, 0.42 ug/uL BSA) are added and the reaction is incubated for 90 min. at 41^oC. The reaction product can be stored at -20^oC until analyzed by hybridization analysis.

NASBA Primers. The following are the NASBA primers utilized in this study. HTLV I tax A: P1= 5'*AATTCTAATACGACTCACTATAGGG*GAGTCGAGGGA TAAGGAACTGTA3'; P2= ACGTGTTTGGAGACTGTGTACAAGG3'. HTLV I tax B: P1= 5'*AATTCTAATACGACTCACTATAGGG*GGAACGGAAGGAGGCCGTTTTGCCA3'; P2= 5'ACGTGAT TTTTTGCCACCCCGGCCA3'. HTLV I Pol: P1= 5'*AATTCTAATACGACTCACTATAGGG*GA GGGAACGTGTCTCTGGGATCGGCTGGA3'; P2= 5'ACCTCCCTTGCTATTCGCCATACCACCCA3'. Small nuclear ribonucleoprotein (snRNP) RNA: P1= 5'*AATTCTAATACGACTCAC TATAGGG*AGAGGCCCGGCATGTGGTGCATAA3'; P2= 5'CAGTATGCCAAGACCGACTCAGA3' (T. Kievits, personal communication) Italicized sequences are the T7 RNA polymerase promoter region. The primers for the HIV-1 Quantitative Assay have been previously described (14).

Detection of NASBA Amplification Products. All NASBA products are detected by hybridization analysis. The oligonucleotide probes used in the hybridization analysis are labeled using three different strategies, depending upon the application. For example, ^{32}P labeling of probes using T4 polynucleotide kinase is used for early stage assay development and optimization. Probes labeled in this way are used in Southern blot analysis of NASBA reaction products, using standard techniques. We also utilize oligonucleotide probes which have been labeled at the 5' end with horse radish peroxidase (HRP) (13). This enzyme linked gel assay (ELGA) involves conventional liquid hybridization of the HRP labeled probe to the NASBA product, followed by resolution through a polyacrylamide gel. The resulting gel is exposed to the appropriate substrate solution and the resulting color reaction indicates the position of the hybridization products in the gel. Finally, an electrochemiluminescence (ECL)-based detection system is used to detect NASBA products (1). This system requires the detector probes to be end labeled with a ruthenium (Ru^{2+}) compound. These Ru^{2+} labeled probes are used in a magnetic bead hybridization capture format, and the hybridization reaction products are measured using the NASBA QR-System, which is an ECL reader (14).

The specific probe sequences used in the detection assays described here are: HTLV I tax A= 5'GTTCGGCCCGCCTACATCGTCACGC3'; HTLV I tax B= 5'AGGGGCCCTAATAATTCTACCCGAA3'; HTLV I pol= 5'ACAATCAACCACCT GAATGTGTTAA3'; snRNP= 5'AGAAGAGGAAGCCCAAGAGCCA3'. The HTLV I capture probe sequence is 5'TGGGACCCCATCGATGGACGCGTTA3'. The HIV-1 probes were previously described (14).

RESULTS

The quantitative NASBA assay for HIV-1 RNA. The basic strategy of this assay has been described elsewhere (14). This assay utilizes the method of Boom et al. (2) for the isolation of nucleic acids from the test sample. However, in order to make this assay quantitative, three internal control calibrator RNA molecules are spiked into the lysate of the clinical sample, prior to the actual extraction phase. These internal control calibrators (or Q RNAs) are identical to a 1.5 kilobase (kB) portion of the wild type (WT) HIV-1 gag region, which encompasses the area targeted by the NASBA primers. However, these Q RNAs have been altered at the 20 base region which is complimentary to the detector probe. In the case of each Q RNA (Qa, Qb, Qc) the detector probe target site has been mutagenized such that the overall A, C, G, and T content is maintained, but the base order is distinct in each case. This will allow for the use of four separate detector probes, one for the WT target, and one for each Q RNA. In each case the probes will hybridize with the same efficiency; however, they will not cross hybridize.

The assay can be used to quantitate the amount of WT RNA in the test sample due to the fact that the concentration of each Q RNA added to the test sample lysate is known (Qa= 10^6 copies, Qb= 10^5 copies, Qc= 10^4 copies). Since the Q RNAs are added immediately after sample lysis, they are coextracted along with the WT RNA present in the sample. Thus, the resulting extract contains all four types of RNA, and the ratio of RNA in this extract is the same as that which was originally established upon the addition of the Q RNAs to the test sample. A small amount of the resulting coextract is then used in a single tube NASBA amplification. The resulting amplification products are therefore present at a ratio to each other which is again equal to the original ratio at the time of lysis.

Actually determining the ratio of WT:Qa:Qb:Qc requires the amount of each amplification product to be quantitated. This is achieved using ruthenium labeled probes for each product and the NASBA QR System. The ruthenium compound allows for the detection of the hybridization products via an ECL reaction (1). The amount of ECL produced in each hybridization can be determined using the NASBA QR system, which is an ECL reader. The QR system utilizes a magnetic bead capture format. Briefly, a biotin labeled capture probe, which will hybridize to each Q RNA and the WT RNA at a conserved site that is independent from the detector probe site, is immobilized onto the surface of a streptavidin coated magnetic bead. This capture probe is included in the four separate detector probe hybridization reactions. These reaction products are then loaded into the QR system, which quantifies the amount of ECL signal obtained in each hybridization reaction. By establishing the ECL signal ratios between WT, Qa, Qb, and Qc, the original WT input quantity can be calculated relative to the Q RNA input quantities using a curve fitting theory algorithm.

Performance Characteristics of the HIV-1 Quantitative Assay. The HIV-1 quantitative assay has been evaluated in a number of studies. For example, it has been shown that the assay is accurate and reproducible within 0.2 log, over a dynamic range that exceeds four orders of magnitude (15). The sensitivity of the assay is a function of the original input sample volume, which can range from as little as 10 uL up to as much as 1.0 mL. The maximum sensitivity determined for the assay is 400 copies of RNA per input volume. It was also shown in this study that performance of the assay is not altered by the anticoagulant used to obtain plasma samples; heparinized plasma can be used as readily as EDTA and citrate preserved samples. Finally, it was demonstrated that the reaction would not be inhibited by large amounts of lipid, hemoglobin, or albumin, which may be present in blood derived samples.

The assay has also been evaluated by means of proficiency panel testing. For example, the AIDS Clinical Trial Group (ACTG) provides proficiency panel samples to laboratories seeking certification for HIV-1 RNA testing in ACTG sponsored clinical trials. The HIV-1 quantitative assay has been evaluated with two of these panels. The first panel consisted of 19 samples at five different input concentrations ranging from 2,400 copies/mL up to 1,500,000 copies/mL. The mean log standard deviation (SD) determined for this panel with the NASBA assay was 0.13; the mean log accuracy was 0.12. In a second proficiency panel evaluation, a total of 20 samples with input quantities above the cutoff for the assay were tested and the mean log SD was 0.09; the mean log accuracy was 0.05 (12). These results indicate that the quantitative HIV-1 NASBA assay is capable of performing with consistently high levels of accuracy and reproducibility over a broad dynamic range.

Applications of the Quantitative HIV-1 NASBA Assay in the Study of AIDS. This assay has been applied in a number of biological studies, designed to evaluate the significance of HIV-1 load. For example, the assay has been used to evaluate therapy efficacy. Kaplan et al. (9) used the assay to demonstrate that patients receiving foscarnet therapy for AIDS related CMV retinitis had marked decreases in their HIV-1 plasma load. In a study of HIV-1 infected patients receiving AZT therapy, it was shown that reduction in load over the first eight weeks of therapy was of greater magnitude in patients who responded clinically to the therapy (8). NASBA derived plasma levels of HIV-1 were also used to determine patient prognosis. In a retrospective study, it was demonstrated that long term nonprogressing patients had significantly lower HIV-1 RNA plasma levels shortly after seroconversion than did progressor patients (6). The NASBA assay was also proven effective for the study of HIV-1 natural infection history in individual patients (3), and in groupings of patients made according to disease status (7). The use of the assay has not been restricted to the analysis of blood. For example, Ginocchio et al. (1995) were able to show that the levels of HIV-1 RNA determined by NASBA were useful in the differential diagnosis of AIDS related encephalopathies.

Qualitative Detection of HTLV I RNA. The NASBA system has also been configured for the detection of HTLV I RNA. Specifically, we designed two primer sets which will amplify the *tax* RNA, and one primer set which targets the *pol* region. These primer sets were evaluated for sensitivity and specificity using RNA obtained from various cell lines. For example, all three primer sets were used in the amplification of 5 ng of total RNA from two HTLV I chronically infected cell lines (MT2, NS-1), an HTLV II chronically infected cell line (MoT), as well as other neoplastic T cell lines (with and without chronic infection by HIV-1). The only RNAs amplified with all three HTLV I primer sets were obtained from MT2 and NS-1. No cross reactivity was observed with RNA from any of the other cell lines. Sensitivity of the three primer sets was determined by attempting NASBA amplification on dilutions of total RNA obtained from MT2 cells. The tax A primer set was able to amplify 5 pg of total RNA from these cells; the tax B and pol primers were able to amplify down to 500 pg and 5 ng, respectively. The difference in sensitivity for the two tax primer sets, which are specific for the same RNA molecule, indicates a primer based component to NASBA sensitivity.

This is most likely due to secondary structure issues which are typical with RNA analytes. Despite the fact that the same number of target molecules are present for the tax A and B primer sets, the two sets of primers may not be able to anneal to their respective targets with equal efficiency.

In addition to utilization of these primer sets on model systems, we also applied the HTLV I primer sets in the analysis of patient samples. In a pilot study, nucleic acids were isolated from patient peripheral blood mononuclear cell (PBMC) cultures that had been viably cryopreserved. There were a total of 15 patient samples available for this study; eight were HTLV I positive, and seven were HTLV II positive. Prior to analysis with the HTLV I primer sets, the integrity of the RNA obtained from each sample was diagnosed using an alternative NASBA assay. This system targets the RNA which encodes the A protein of the U1 small nuclear ribonucleoprotein (snRNP) particle (5). The U1 snRNP particles are involved in the mRNA splicing process. Transcripts encoding the component proteins of the particle are expressed in all human cells. The protein A transcript is a low abundance mRNA target. As such, the detection by NASBA of this target would suggest that isolation of RNA from the cellular source was successful and that there was proper handling of the sample. When this assay was applied to the 15 patient PBMC extracts, RNA was detected in 12 of the samples (7 HTLV I patients, 5 HTLV II patients). Therefore, only these 12 samples were subjected to subsequent NASBA analysis.

During the early stages of NASBA assay development, amplification reaction products are detected using ^{32}P labeled probes. Figure 1 depicts the results obtained from such an analysis of the seven snRNP+/HTLV I+ patient PBMC extracts analyzed with the tax primer sets. In the analysis with the tax A primer set, all of the patient samples were positive, although sample F is only weakly positive. In the tax B primer set analysis, all of the samples are again positive, except for sample F which was negative. Thus, the greater sensitivity of the tax A primer set demonstrated during the analysis of the cell line model systems is apparently maintained in the analysis of patient samples.

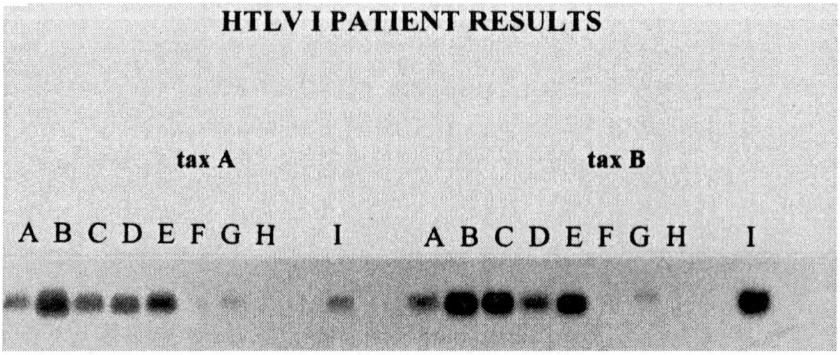

FIGURE 1
HTLV I patient results. Nucleic acid extracts from seven HTLV I infected patients (lanes A through G) were analyzed by NASBA using the tax A and tax B primer sets. The amplification reaction products were resolved through an agarose gel and transferred to nylon membranes using standard techniques. The membranes were then hybridized with ^{32}P labeled probes specific for each NASBA reaction product. After washings, the membranes were exposed to film in autoradiography. Lane H: water only negative control; lane I: MT2 cell line positive control RNA.

Importantly, none of the HTLV II patient samples were positive in the analysis with both HTLV I tax gene primer sets (data not shown). We also analyzed the tax A NASBA products using a nonisotopic detection system, called ELGA (enzyme linked gel assay). The ELGA system utilizes a horse radish peroxidase labeled probe in liquid hybridization format (13). Results from this analysis are provided in Figure 2. Here, results from HTLV I and HTLV II positive patient samples are presented. Results obtained with the ELGA system were identical to those generated with the isotopic detection system.

In addition to the gel based detection systems described above, we also adapted the ECL system for detection of the tax A NASBA product. This required the development of a capture probe which is specific for the tax A NASBA product, at a position which is distinct from the detector probe site. This capture probe was immobilized onto the surface of a magnetic bead by means of a biotin-streptavidin association. After incubating the immobilized capture probe with the

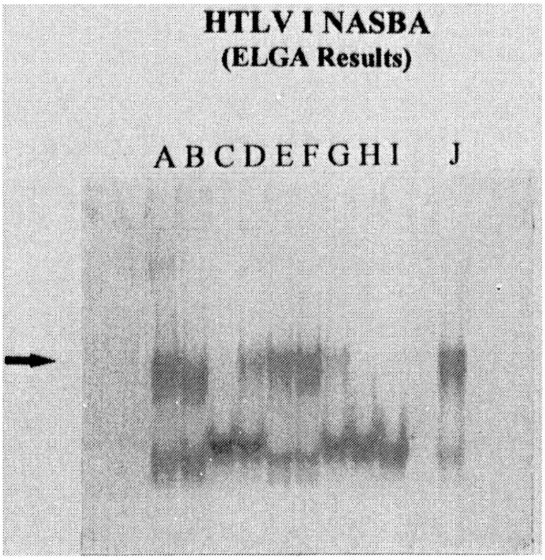

FIGURE 2
HTLV I tax A NASBA ELGA results. Amplification reaction products generated with the HTLV I tax A primer set were analyzed using the colorimetric gel based ELGA detection system. The arrow indicates the position of the HRP labeled probe bound to amplification reaction product in the gel. The lower molecular weight band represents free HRP labeled probe. HTLV I positive samples: lanes A, B, D, E, F, G; HTLV II positive samples: lanes C H; water only negative control: lane I; MT2 cell line positive control RNA: lane J.

tax A NASBA product and the ruthenium labeled detector probe, the hybridization reaction product is loaded into the NASBA QR system. Results from this analysis are provided in Figure 3. The assay negative (AN) consists of immobilized capture probe, detector probe, and water (i.e., no NASBA product). Therefore the ECL signal obtained from the assay negative can serve as an approximation for the background signal produced in this system. Samples A through G all contain HTLV I positive patient PBMC extracts; samples H through L are HTLV II positive patient extracts. In the case of each HTLV I positive patient, the resulting ECL signals exceed that of the assay negative by a factor of at least six (sample F), and by as much as a factor of 1,600 (sample D). Interestingly, the sample F depicted in Figure 3 with the lowest ECL signal among the HTLV I patients, is the same amplification product as lane F in Figure 1. This comparative analysis demonstrates the power of

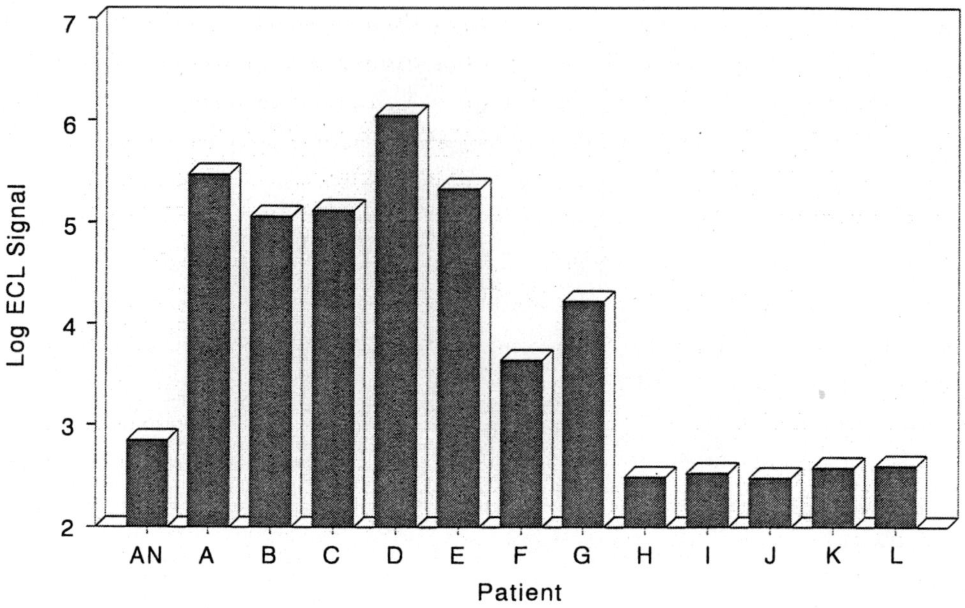

FIGURE 3
ECL detection of HTLV I tax A NASBA products. HTLV I tax A NASBA products were hybridized to a capture probe (which was immobilized onto the surface of a magnetic bead) and a ruthenium labeled detector probe. This product was loaded into the NASBA QR system (Organon Teknika), which quantifies the resulting ECL signal. AN= assay negative; HTLV I positive samples= A through G; HTLV II positive samples= H through L.

the ECL detection methodology. In the case of the autoradiography signal produced after hybridization with a ^{32}P labeled probe, the signal is only slightly above background level, making an interpretation of this result somewhat difficult. However, when the same amplification product is subjected to the ECL detection procedure, its signal is six times greater than the background, clearly establishing it as a positive sample. Further, the specificity of the system is maintained using the ECL detection procedure; all five HTLV II patients (lanes H through L) were again negative.

DISCUSSION

NASBA is a total system, comprised of three separate technologies; nucleic acid isolation, isothermal amplification, and hybridization based detection. The nucleic acid isolation procedure can be applied to a wide range of sample types and yields highly purified nucleic acids which are suitable for enzymatic amplification. The actual amplification procedure is an isothermal transcription based system which involves the coordinated activities of three enzymes and two oligonucleotide primers. Amplification by NASBA is most appropriate for RNA analytes due to the direct incorporation of the reverse transcription process into the amplification reaction. Detection of NASBA products is possible through several methods, depending upon the actual application. Qualitative detection of NASBA amplification products is possible with gel based procedures (e.g., Southern blot, liquid hybridization) using isotopic and colorimetric systems. Quantitative detection of NASBA products makes use of ECL chemistry and an automated reader system.

The appropriateness of the NASBA system for RNA analytes has led to its extensive development for the detection and quantitation of retrovirus RNA. The HIV-1 quantitative NASBA assay makes use of three internal control RNA calibrators, which serve as system controls from the earliest point in the assay. By spiking known quantities of these calibrator RNAs into a sample lysate, they are coextracted and coamplified along with any wild type RNA that is present in the sample. At the detection stage, the coamplificate is divided and subjected to hybridization analysis with four different probes, one for each calibrator and the wild type. By establishing the ECL signal ratio between each calibrator RNA and the wild type RNA, the original wild type input can be calculated. A NASBA based assay for the qualitative detection of HTLV I RNA has also been developed. In this system, nucleic acid isolation and amplification are conducted in the same manner as in the quantitative configuration of the system. However, multiple methods can be applied for detection, including Southern blot analysis with a ^{32}P labeled probe, liquid hybridization using a horse radish peroxidase labeled probe (i.e., ELGA), or the ECL based system.

Evaluation of both of these assays on clinical samples has revealed that the NASBA system is a highly useful tool for RNA diagnostics. The capacity to develop the technology for quantitative and qualitative applications makes it particularly relevant for the study of Retroviruses. Further, it is clear that the NASBA system can be applied to systems outside of virology, including other infectious diseases, oncology, genetic screening, and cell biology.

ACKNOWLEDGMENTS

The authors wish to thank Tim Kievits and Bob van Gemen of Organon Teknika, for their kind provision of reagents and technical advice.

REFERENCES

1. G. Blackburn, H. Shah, J. Kenten, et al., Clin. Chem. 37:1534, 1991
2. R. Boom, C. Sol, C. Jensen, et al., J. Clin. Micro. 28:495, 1990
3. M. Dalod, S. Fiorentino, C. Delamore, et al., AIDS Res. Hum. Retro. 12:497, 1996
4. C. Ginocchio, J. Romano, M. Kaplan, et al., 95th General Meeting of the Amer. Soc. Micro., Washington., D.C., Jan. 29-Feb. 2, 1995
5. M. Green, Ann. Rev. Cell Biol. 7:559, 1991
6. E. Hogervorst, S. Jurriaans, F. de Wolf, et al., J. Inf. Dis. 171:811, 1995
7. S. Jurriaans, B. van Gemen, G. Weverling, et al., Virology 204:223, 1994
8. S. Jurriaans, G. Weverling, J. Goudsmit, et al., AIDS Res. Hum. Retro. 11:473, 1995
9. M. Kaplan, C. Ginocchio, J. Romano, AIDS Res. Hum. Retro. 11(s1):S83, 1995
10. A. Katz, C. Olson, A. Raffo, et al., Urology 43:765, 1994
11. T. Kievits, B. van Gemen, D. van Strijp, et al., J. Virol. Meth. 35:273, 1991
12. J. Romano, P. van de Wiel, S. Geiger, submitted.
13. G. van der Vliet, R. Schukkink, B. van Gemen, et al., J. Gen. Microbiol. 139:2423, 1993
14. B. van Gemen, R. van Beuningen, A. Nabbe, et al., J. Virol Meth. 49:157, 1994
15. B. van Gemen, P. van Wiel, R. van Beuningen, et al., PCR Meth. Appl. 4:S177, 1995

CONCERTED USE OF IMMUNOLOGIC AND ULTRASTRUCTURAL ANALYSES IN DIAGNOSTIC MEDICINE: IMMUNOELECTRON MICROSCOPY AND CORRELATIVE MICROSCOPY

S. E. Miller[1, 2] and D. N. Howell[2, 3]
Departments of Microbiology[1] and Pathology[2]
Duke University Medical Center, Durham, North Carolina 27710
Department of Pathology[3]
VA Medical Center, Durham, North Carolina 27705

ABSTRACT

Electron microscopy (EM) is a valuable tool in diagnostic medicine, and in some cases, can be enhanced by immunological methods. A major medical application of EM, diagnostic virology, can frequently be augmented by employment of immunological reagents. Three immunoelectron microscopy (IEM) methods, aggregation, coating, and gold labeling, provide means for serotyping viruses; aggregation by antibody can also be used to concentrate viruses in dilute suspension or to serotype them.

As a research tool, IEM can be useful in studying the relationship of various pathogen proteins to the infected cells or tissues. Delineating the subcellular location of viral components can yield information about how virions are constructed, and hence, suggest methods and compounds for inhibiting that process. Conversely, labeling virus-infected cells with antibodies against various cell receptors and proteins can yield information about the association of the proteins with budding virions. Another research example is the identification by immunological staining of virus-infected cells for subsequent ultrastructural identification of the specific cell type involved.

Electron microscopy and immunolabeling methods are also useful in the diagnosis of immune complex disorders, including various forms of postinfectious immune complex glomerulonephritis. Precise analysis of immune complex deposits can be accomplished by using EM to pinpoint their location and immunohistology to probe their composition.

Finally, a variety of optical microscopic techniques, including some involving immunofluorescent labeling, can be used to identify areas of interest in inhomogeneous tissues for further study by EM.

INTRODUCTION

Electron microscopy (EM) is a powerful tool for the investigation of infectious processes. It is widely applied in the diagnosis of viral illnesses (1, 2), but can be used to detect infections with other agents such as bacteria, mycoplasmas, and fungi. Electron microscopy is also of crucial importance in the analysis of certain indirect systemic consequences of infection, such as postinfectious immune complex glomerulonephritis.

In each of these applications, EM provides ultrastructural resolution necessary for direct visualization of the pathologic process. In obtaining this resolution, however, EM places severe limitations on the amount of tissue that can be examined; focal pathologic processes can easily be missed by routine ultrastructural examination. Similarly, dilute particulate samples such as viruses in suspension may go undetected. Conventional EM also provides little biochemical information about the specimen under study; it cannot, for instance, identify the serotype of an infectious virus particle or the subclasses of immunoglobulin molecules in an immune complex deposit.

Several of the drawbacks of EM are overcome by immunohistologic methods, in which tissues are labeled with antibodies specific for exogenous and/or endogenous antigens (e.g., viral proteins, immunoglobulin isotypes) (3). When applied at the light microscopic level, these methods can be used to search large tissue specimens for focal pathologic processes. With properly selected antibody reagents, they can also provide very specific biochemical information about infectious agents and the immune response mounted against them.

The advantages of immunohistology, conversely, are offset somewhat by suboptimal resolution. For example, immunofluorescence analysis for immunoglobulin isotypes can provide precise information regarding the composition of glomerular immune complex deposits, but only an approximate idea of their distribution. Immunohistologic approaches in general are also limited by availability and selection of antibody reagents.

The respective weaknesses of EM and immunohistology can be overcome in many cases by combined use of the two investigative modalities. In some instances, the combination may simply involve separate, parallel or sequential examinations of a specimen by both methods, a technique referred to as "correlative microscopy." In other applications, immune reagents can be used to stain tissues and aggregate or concentrate particulate specimens for direct examination by EM; these methods are collectively designated "immunoelectron microscopy" (IEM). This presentation will describe applications of both types of microscopy to the diagnosis of infectious diseases and their complications.

MATERIALS AND METHODS

Fluorescence Microscopy (4)

Tissue was embedded in gelatin, snap frozen in 2-methylbutane in a liquid nitrogen bath, and stored at -70°C. Frozen sections were cut at 4 µm, air dried, fixed in acetone, and stained by the direct immunofluorescence method using FITC-conjugated antibodies against human IgG, IgM, IgA, C3, C1q, fibrinogen (Cappel Co., Durham, NC), albumin (Bionetics, Kensington, MD), and immunoglobulin kappa and lambda light chains (Dako Corp., Carpinteria, CA). Immunofluorescence microscopy was performed using an Olympus BH-2 microscope equipped with a BH2-RFC epifluorescence attachment (Olympus Corp., Lake Success, NY).

Confocal Microscopy (5)

Tissue was fixed in 2% glutaraldehyde in 0.1 M cacodylate buffer, pH 7.4, containing 3.4% sucrose. Following fixation, 100-µm sections were cut on a vibrating microtome (EMS, Fort Washington, PA). The sections were stained by immersion in 50 µg/ml propidium iodide in distilled H_2O for 3 min at 25°C, then examined using a slit-scanning confocal microscope (InSIGHT Plus, Meridian Instruments, Inc., Okemos, MI) equipped with a green excitation filter cube. Areas of interest were excised using a razor blade and processed for electron microscopy as described below, with embedment in flat molds. Care was taken to insure that the surface of the tissue examined by confocal microscopy was embedded parallel to and facing the end of the block to be sectioned (6).

Immunolabeling for electron microscopy

Small round viruses were obtained from stool by water extraction and low speed centrifugation to remove large debris (2). Virus aggregation was performed by adding monoclonal anti-poliovirus type 1, 2, or 3 antiserum at a final dilution of 1:1000 to the virus suspension, incubating 1 hr at 37°C, and pelleting the aggregates at 17,000 x g for 1.5 hr (7). A grid with a support film was then placed onto a drop of the pellet suspension, drained, and negatively stained as described below.

Rotaviruses extracted from stool were adhered to grids with support films and then labeled by adding specific rabbit anti-rotavirus (either type A or B) serum, washing, and adding goat anti-rabbit serum conjugated to 10-nm colloidal gold. Immunological methods for virus identification have been reviewed by Hayat and Miller (8).

Staining of HIV-infected cells was performed on live cells at 4°C with mouse monoclonal antibody against CD44, followed by goat anti-mouse serum conjugated to 10-nm colloidal gold; cells were washed 3 times in buffer after each antibody. Cells were then fixed and embedded for EM routinely as described below.

Staining of ultrathin cryosections for localization of internal antigens (9) was performed directly on grids using drops of primary antiviral antibody: in the case illustrated, mouse monoclonal antibody against an Ebola virus nucleocapsid protein. Secondary labeling was with rabbit anti-mouse serum tagged with 10-nm colloidal gold. Other primary antibodies were detected with either protein A- or protein G-gold. Controls were an unrelated antibody of the same type and uninfected cells.

Specimen preparation and Electron Microscopy

Negative staining of whole virus preparations on Formvar and carbon-coated grids was with uranyl acetate. Hayat and Miller have reviewed negative stains and procedures (8). Routine processing of tissue included glutaraldehyde, osmium, and uranyl acetate fixation, followed by dehydration in graded ethanols and embedment in Spurr resin (10). Ultrathin epoxy sections were cut on a Reichert-Jung Ultracut E ultramicrotome (Leica, Deerfield, IL) and post-stained with uranyl acetate and lead citrate. Ultrathin frozen sections were cut in an FC 4E cryochamber attachment for the ultramicrotome. Electron microscopy was performed on a Philips EM300 electron microscope (Philips Electronic Instruments, Mahwah, NJ).

RESULTS

Immunoelectron microscopy

Virus serotyping. Antiserum against poliovirus type 3 aggregated the small viruses isolated from stool of an immune deficient child (Fig. 1A), while antisera to types 1 (Fig. 1B) and 2 did not. Rotaviruses isolated from cases of gastroenteritis in North Carolina reacted with anti-rotavirus type A and were labeled by colloidal gold-conjugated secondary antibody (Fig. 2A). Antibody against rotavirus type B did not recognize these viruses (Fig. 2B).

Extracellular labeling. Labeling HIV-infected cells with anti CD44, followed by goat anti-mouse antiserum conjugated with colloidal gold, showed that the virus particles are associated with the antigen and carry it out as they are released from the cells (Fig. 3).

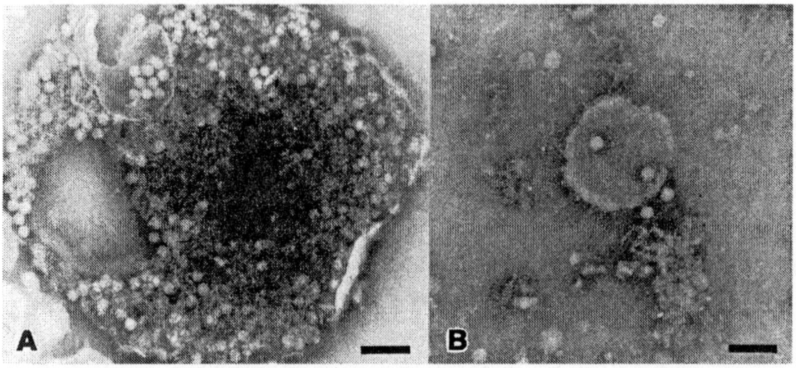

FIGURE 1
Serotyping by immunoaggregation. A. A small round virus from stool reacted with anti-poliovirus type 3 antiserum; the antibodies aggregated the virions into large clumps. B. The same virus reacted with anti-poliovirus type 1 antiserum; clumps were not formed. Bars represent 100 nm.

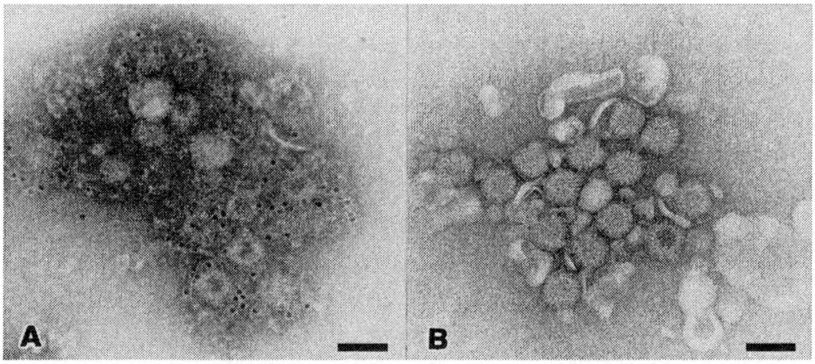

FIGURE 2
Serotyping by immunolabeling. A. Rotavirus from stool was reacted with antirotavirus type A antiserum, followed by gold-conjugated secondary antiserum. Black dots indicate the positive reaction. B. The primary antibody used was antiserum against group B rotavirus. Bars represent 100 nm.

Intracellular localization. Cultured cells were stained with a monoclonal antibody against an Ebola virus nucleocapsid protein; the reaction was identified by gold-labeled secondary antiserum. In preliminary experiments, high background staining was observed in inappropriate subcellular locations; this was eliminated by absorption of the primary antibody preparation with uninfected cells. When infected cells were stained with the absorbed primary antibody (Fig. 4A), virus factories inside the cells were labeled, as well as budded virions at the surface. An unrelated mouse primary antibody used as a control was nonreactive (Fig. 4B). Uninfected cells were heavily labeled by unabsorbed primary antibody (Fig. 4C), but no staining was present when absorbed primary was used (data not shown).

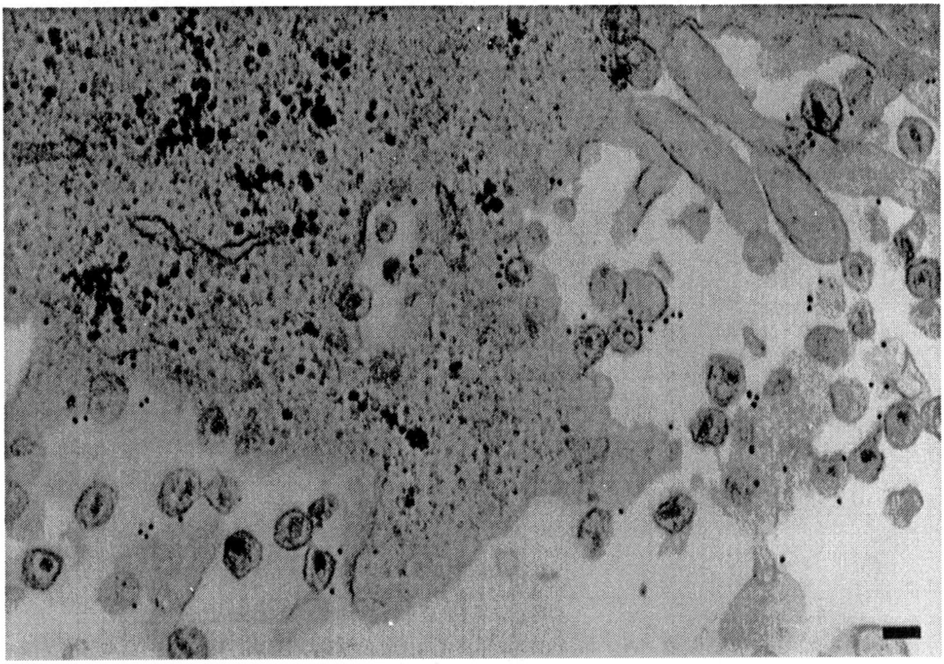

FIGURE 3
Extracellular antigen detection. Antiserum against the cell surface receptor CD44 reacts not only with the surfaces of HIV-infected cells but also with the virions, and frequently is seen at the necks of budding particles. Cells provided by Dr. Barton Haynes. Bar represents 100 nm.

Correlative Microscopy

Immune complex diseases. Figure 5 includes three micrographs from a kidney biopsy performed on a patient with hepatitis C virus (HCV) infection, cirrhosis, and progressive renal failure. Light microscopy (Fig. 5A) demonstrated diffuse proliferation of cells within glomerular tufts. By immunofluorescence microscopy, granular immune complex deposits containing IgG, IgM, and immunoglobulin kappa (Fig. 5B) and lambda light chains were detected within glomeruli. Though some of were located in capillary loops, their exact location with respect to the capillary basement membranes could not be ascertained. This problem was solved by EM (Fig. 5C), which demonstrated immune complex deposits (arrowheads) on the inner ("subendothelial") faces of the basement membranes.

Selection of tissue samples for EM. Correlative microscopy can be used to identify focal infectious lesions for subsequent examination by EM. This principle was applied to the analysis of autopsy tissue from a patient with rare renal granulomas. Using a confocal microscope, a granuloma was located in a vibrating microtome section of glutaraldehyde-fixed tissue stained with propidium iodide (Fig. 6A; *); an adjacent vascular space (v) and a portion of a glomerulus (g) are also visible in the micrograph. The area of interest was excised and embedded for EM. Light microscopic examination of a toluidine blue-stained thick epoxy section (Fig. 6B) confirmed the

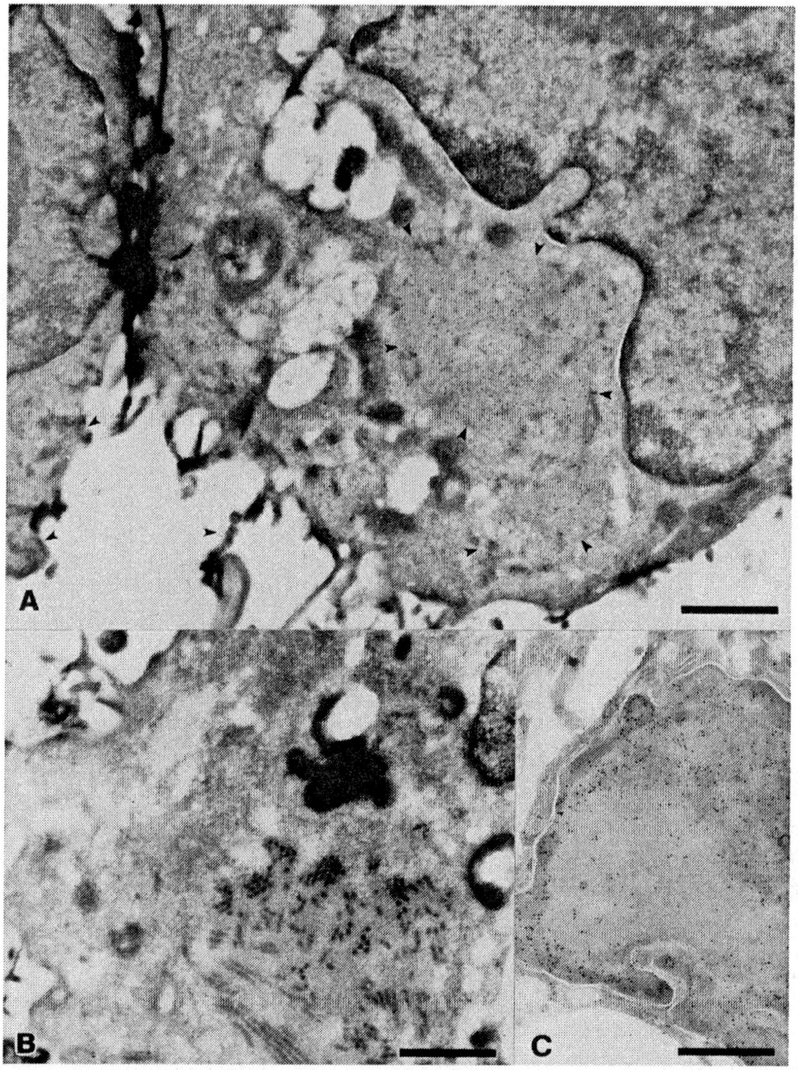

FIGURE 4
Intracellular antigen labeling. Ultrathin cryosections of Ebola virus-infected cells were labeled with a monoclonal antibody against one of the nucleocapsid proteins. A. Antibody absorbed with uninfected cells shows labeling over the virus factories in the cytoplasm of cells as well as a little labeling over virions; only virions cut so that the internal structure is available to the antibody are stained. B. Control infected cells reacted with an unrelated antibody are not stained. C. Infected cells reacted with unabsorbed antibody have high background over both the cytoplasm and nucleus of the cell. Bars represent 1 μm.

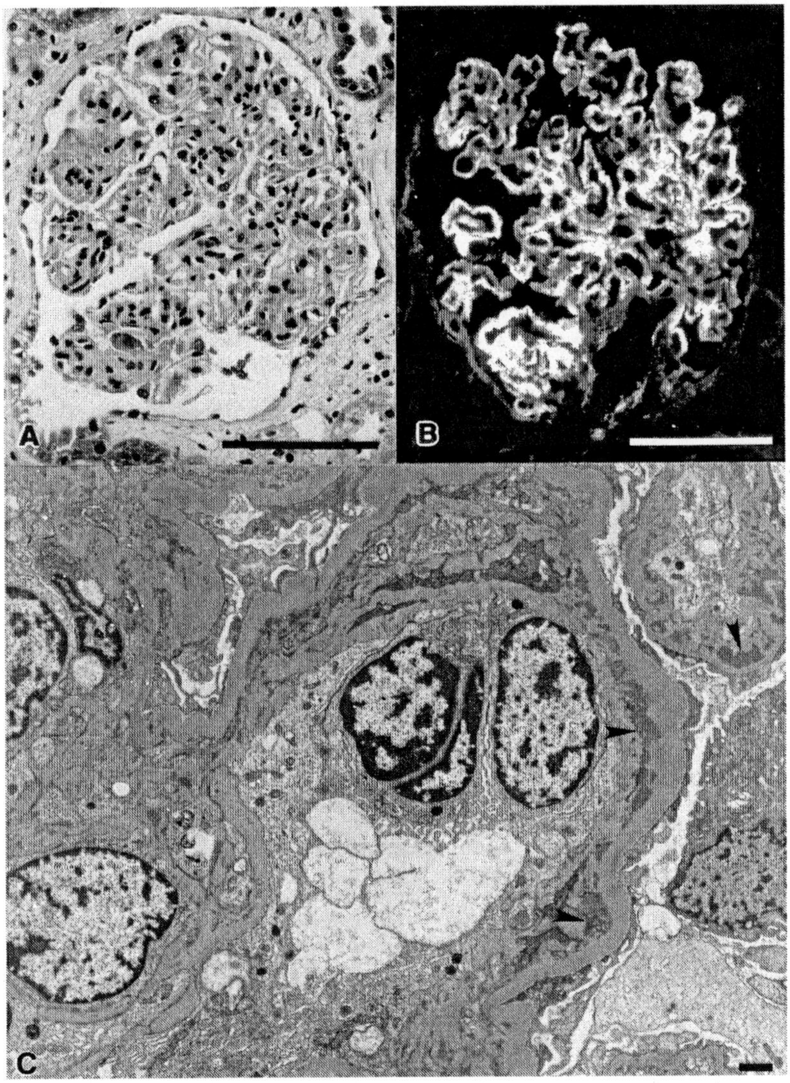

FIGURE 5
Correlative microscopy: Immune complex detection. A. Light micrograph of an H & E-stained paraffin section of a glomerulus from a patient with HCV infection demonstrating increased cellularity. B. An immunofluorescence micrograph of a glomerulus stained for immunoglobulin kappa light chain shows granular immune complex deposits in capillary loops. Bars in A and B represent 300 μm. C. The exact location of the deposits (arrowheads) was demonstrated by electron microscopy to be on the inner faces of capillary loop basement membranes. Bar represents 10 μm.

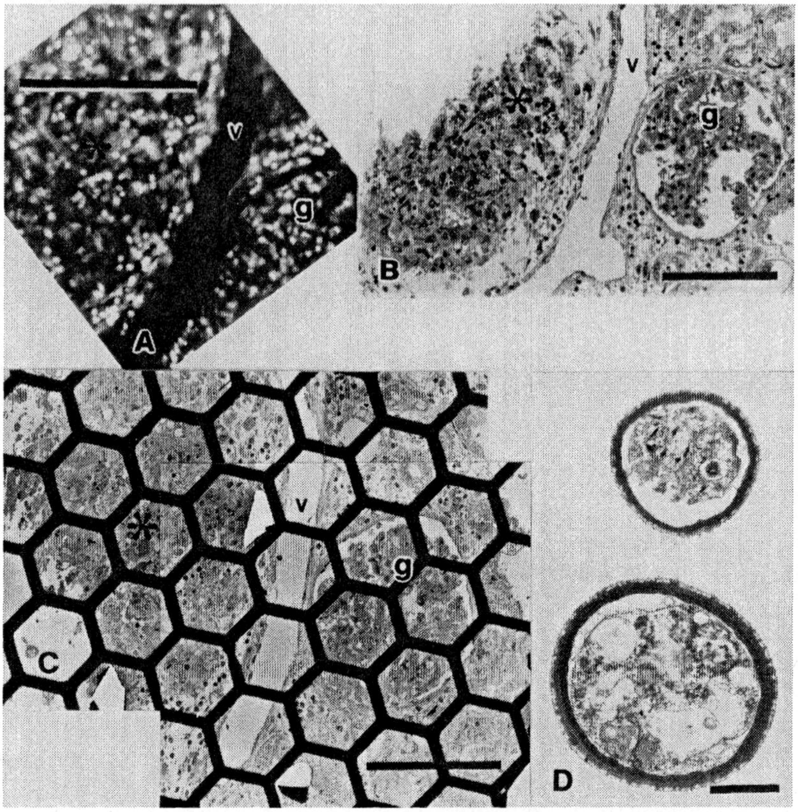

FIGURE 6
Correlative microscopy: Confocal microscopy selection of tissue for EM exam. A. Confocal micrograph of a vibrating microtome section of kidney showing a granuloma (*), a vessel (v) and a portion of a glomerulus (g). B. Light micrograph of a toluidine blue-stained thick epoxy section of the same area shown in A. C. Very low magnification EM of a thin section from the same area showing the same structures. Bars in A-C represent 100 μm. D. High magnification of yeasts seen in the granuloma. Bar represents 1 μm.

presence of the granuloma and adjacent structures; and a very low magnification electron micrograph of thin sections produced following the thick section also shows the granuloma (Fig. 6C). At higher magnification, EM revealed rare yeasts (Fig. 6D).

DISCUSSION

Virus serotyping can be performed with specific antiserum if identification has been narrowed to allow selection of the proper reagents and if the reagents are available. In the example shown here, small round viruses seen in the stool of a patient with severe combined immunodeficiency were identified as one of the vaccine strains of poliovirus by clumping with antivirus antiserum.

Similarly, antisera against different types of rotaviruses followed by colloidal gold secondary antibody permitted specific identification of North Carolina strains to be type A, and not the more virulent Asian form, type B. Such reactions permit epidemiological studies of various virus infections.

Labeling of extracellular antigens such as the CD44 receptor shown here permit study of the cell surface antigens as well as the relationship of surface proteins to other processes, such as budding viruses or attachment sites to other cells.

Concerted use of EM and immunohistochemistry is a mainstay in the diagnosis of immune complex diseases. This is particularly true for the study of renal biopsies, where EM is routinely performed in tandem with immunofluorescence microscopy using antisera against immunoglobulins and complement components. Immunofluorescence provides information about the composition of immune complexes, while EM pinpoints their location.

Complexes on the subendothelial faces of capillary loop basement membranes are seen frequently in patients with glomerulonephritis associated with HCV infection (11). In contrast, glomerular disease associated with several other forms of infection (e.g., post-streptococcal glomerulonephritis) typically involves deposition of immune complexes on the outer ("subepithelial") aspect of the basement membrane.

Several investigators have obtained ultrastructural and biochemical information about immune complexes simultaneously by employing immunogold labeling and immunoelectron microscopy (12, 13, 14). As of yet, however, this approach has been confined to the research laboratory, and has not been applied as a routine diagnostic tool.

Tissue blocks for routine examination by EM are limited to a size of approximately 1 mm^3. Since many pathologic processes, including infections, are very focal, they may not be represented adequately in small random samples taken for EM study from a larger tissue specimen. This problem can be solved using correlative microscopy. We have recently developed a technique in which large tissue samples are examined by confocal laser scanning microscopy to identify areas with suspicious features (e.g., viral cytopathic changes, necrosis, inflammation) for subsequent analysis by EM. Confocal microscopes allow direct examination of relatively thick tissue specimens by producing "optical sections" of thin planes within the tissue.

The confocal microscopy step can be tailored to a variety of fluorescent probes. In research settings where infection with a specific pathogen is known or suspected, infectious foci can be identified by staining with fluorescently-tagged antibodies for subsequent ultrastructural study. The technique can also be generalized to other diagnostic pathology settings, including ultrastructural analysis of small foci of neoplastic tissue.

ACKNOWLEDGMENTS

The authors are grateful for technical assistance from Steve Conlon, Mike Hale, Susan Hester, Ben Hopkins, and Susan Reeves.

REFERENCES

1. S.E. Miller. J Electron Microsc Tech 4, 265-301 (1986).
2. S.E. Miller, in Diagnostic Procedures for Viral, Rickettsial and Chlamydial Infections, E.H.

Lennette, D.A. Lennette, and E.T. Lennette, eds, American Public Health Assoc, Washington, DC, (1995) pp. 35-76.

3. R.B. Colvin, A.K. Bhan, and R.T. McCluskey, eds, Diagnostic immunopathology. Raven Press, New York.
4. A. Kawamura, and Y. Aoyama, eds, Immunofluorescence in Medical Science. Springer-Verlag, New York, (1983).
5. J.B. Pawley, Handbook of Biological Confocal Microscopy, 2nd ed. Plenum Press, New York, (1995).
6. S.E. Miller, D.N. Howell, and R.M. Levenson. Proc Southeastern Micros Soc 17, 32 (1995).
7. A.Z. Kapikian, J.L. Dienstag, and R.J. Purcell, in Manual of Clinical Immunology, N.R. Rose, and J. Friedman, eds, American Society for Microbiology, Washington, DC, (1976) pp. 467-480.
8. M.A. Hayat and S.E. Miller, Negative Staining: Applications and Methods. McGraw-Hill, New York, (1990).
9. G. Griffiths, Fine Structure Immunocytochemistry. Springer-Verlag, New York, (1993).
10. J.J. Bozzola and L.D. Russell. Electron Microscopy. Jones and Bartlett Publishers, Inc, Boston, MA (1992).
11. R.J. Johnson, D.R. Gretch, H. Yamabe, et al. N Engl J Med 328, 465-70 (1993).
12. M.M. Silver, S.A. Hearn, S. Ritchie, et al. Am J Pathol 122, 17-27 (1987).
13. P.S. Smith. Pathology 20, 32-37 (1988).
14. D.C. Matthews, J.A. Bell, J.P. Dowling, and P.S. Kincaid-Smith. 8, 43-48 (1985).

ADSORPTION-INDUCED ANTIGENIC CHANGES AND THEIR SIGNIFICANCE IN ELISA AND IMMUNOLOGICAL DISORDERS

J. E. Butler, P. Navarro and J. Sun
Department of Microbiology
University of Iowa
Iowa City, IA 52242

ABSTRACT

The functional properties of ^{125}I-labeled antibodies and antigens adsorbed on polystyrene and silicone were compared to their counterparts immobilized by non-adsorptive methods. Less than 20% of polyclonal (pAb) and 1-2% of monoclonal (mAb) capture antibody equivalents remained functional after adsorption as a monolayer. Survivability circa doubled or was totally rescued, when the same antibodies were immobilized via a streptavidin bridge or by using a first stage polyclonal antiglobulin capture antibody, respectively. Similarly, the antigenicity of bovine IgGs for pAb and mAb anti-IgGs was highest when the IgGs were immobilized via a streptavidin bridge or when secondarily adsorbed to an albumin monolayer. IgGs in these configurations were significantly more antigenic than when directly adsorbed on polystyrene or a silicone elastomer. Similar activity was seen after adsorption on polystyrene or silicone. Interestingly, these IgGs were equally antigenic when denatured and subsequently adsorbed in 6M guanidine-HCl versus adsorption in PBS without prior denaturation. Although many of the above finding on antibodies and antigens could be explained by the greater accessibility of antigenic epitopes or antibody binding sites when molecules are immobilized by some type of underlying molecular layer, we also show that certain mAb and pAbs preferentially recognized allotopes on IgG2a when IgG2a was adsorbed. Furthermore, such antigenicity was highest when IgG2a was adsorbed at low, sub-monolayer concentrations.

Finally, we show that differences in antigenicity need not be related to the method of immobilization, but can also result from differences in the microenvironment of the epitope. This was demonstrated using a filamentous phage clone specific for fluorescein (FLU). This clone recognizes the fluorescein hapten differently depending on the carrier protein used and the method of conjugation.

Data presented in this report indicate that antibodies and antigens adsorbed on hydrophobic polymers undergo changes in their functional properties. Data suggest that both changes in conformation and the accessibility of antigen epitopes or antibody binding sites, most likely occur. Especially in the case of the latter, the functional concentration may be 1-2 orders of magnitude lower than the antibody protein concentration. These observations have implications for immunodiagnostics and emphasize the need to determine the specificity of an antibody in the assay in which it is employed and to make no assumptions about the behavior of solid-phase antigens and antibodies from their behavior in solution.

Our studies are also relevant to the use of silicone medical prostheses. The antigenicity of IgGs adsorbed on silicone as a multilayer (secondary layer) is much higher than when directly adsorbed. Since such surfaces would be exposed to very high protein concentrations *in vivo*, multilayers not a monolayer, would be expected. Thus it would seem from these studies that host protein adsorbed on silicone would be expressed to the immune system at the surface of multilayers. This being the case, it seems unlikely that the adsorption of host protein *in vivo* would generate new epitopes against which the host's immune system could respond and subsequently initiate an autoimmune syndrome.

INTRODUCTION

The interaction of antibodies and antigens is a natural *in vivo* event which is part of the host's protective immune response. The specificity and high affinity[1] of these interactions have made them valuable for *in vitro* detection and quantification of antigenic material, and also for *in vitro* measurements of the immune response. Since the discovery and use of immunoassays in the late 19th century, separation of the antigen-antibody complex from the remaining milieu has been fundamental to the technology. Since some complexes form visible precipitates or agglutinate cells, separation was straightforward. However immunoassays based on agglutination and precipitation are restricted to certain antibodies and types of antigens and especially precipitin assays, are not very sensitive. Substantial improvement in sensitivity came about with the development of radioimmunoassay (RIA; 1), a competitive-based assay which also allowed low molecular weight (haptenic) antigens to be quantified. Nevertheless, separation of the antigen-antibody complex from the remaining milieu remained a technical hurdle and remained "the slowest ship in the immunoassay convoy". Thus, the concept of immobilizing one reactant on a synthetic solid-phase became popular. While many ingenious methods of immobilization were used, by far the simplest and most convenient was based on the observation that proteins (including antigens and antibodies) readily adsorb to hydrophobic synthetic polymers in nearly non-dissociable fashion (2). Thus was born what would become known as solid-phase immunoassay (SPI) with detection systems based on radioactivity (RIA, IRMA), colored and chemiluminescent enzymatic products (ELISA) and fluorescence.

SPIs have a number of notable characteristics. First, they permit a very large amount of reactant to be localized in a small area, probably within <100 Å from the surface. This theoretically favors the binding of low affinity antibodies or antigens and for all interaction, the slow rate of dissociation leads to high affinity[1] constants between the soluble reactant and the immobilized reactant. These types of interfacial interactions generally have equilibrium constants 2-log greater than their solution phase counterparts (3, 4). This feature has made it possible to conduct multi-step immunoassays, especially exemplified by ELISAs, without significant dissociation of the reactants during the process. Second, the use of plasticware-type solid phases, i.e. microtiter wells and large spheres, introduced a degree of diffusion dependence not associated with fluid-phase interactions. Finally, the widely-used method of

---

[1]The term affinity and avidity are used interchangeable although recognizing that most solid-phase interaction are multivalent and their higher equilibrium constants reflect such multivalent interaction and should be more correctly expressed as avidity constants.

simple adsorption of proteins on hydrophobic polymers can give rise to conformational change, i.e. adsorption-induced conformational change (AICC; 5).

In data presented here, we argue that changes in the antigenicity of adsorbed proteins and antibodies, can at least in part, be ascribed to AICC. Most of the data presented will focus on antigenic changes and their significance in the development of immunoassays and their potential for producing immunological disorders resulting from the use of hydrophobic medical prostheses.

METHODS AND RESULTS

The Loss of Activity by Adsorbed Capture Antibodies (CAbs)

Two-site or so-called sandwich SPIs as well as certain competitive forms of ELISAs and RIAs, require that capture antibodies (CAb) be immobilized on a solid-phase. Figure 1 illustrates two configurations for immobilizing CAbs, direct adsorption or immobilization via a streptavidin bridge (6). CAbs immobilized by the latter method are >5-fold more active in their ability to capture antigen than those directly adsorbed on polystyrene (Fig. 1). Figure 2 shows that on a quantitative basis, <20% of adsorbed polyclonal CAb and as low as 1-2% of adsorbed monoclonal CAbs, remain functional (expressed as the proportion of CAbeqv) when adsorbed on polystyrene microtiter wells. However, when the same CAbs are immobilized via a streptavidin bridge or by first using an adsorbed polyclonal to "capture the anti-FLU capture antibodies", activity is rescued; even total rescue (200 CAbeqv means both binding sites are active) can be achieved using the latter method.

These studies provide quantitative evidence of the magnitude of CAb activity loss resulting from adsorption. Although it is unlikely that the magnitude of this effect can be totally ascribed to steric hindrance, rescue of activity appears correlated with positioning the CAb at some distance from the surface (Fig. 1).

Changes in Antigenicity of IgGs Adsorbed on Polystyrene or on a Silicone Elastomer

Figure 3 illustrates three different methods for the immobilization of IgG on polystyrene and on a silicone elastomer (PEP)[2]. The first involves simple adsorption, and may be done at very low (sub-saturating) concentration or at a concentration which result in a monolayer of adsorbed IgG (Fig. 3A; 7, 8). Alternatively, IgG can be biotinylated and non-adsorptively immobilized on commercially available streptavidin microtiter wells (SA-PS; Wallac, Finland; Fig. 3B). Finally, IgG can be adsorbed on a pre-formed monolayer of serum albumin by an apparent protein-protein interaction (Fig. 3C). In every case, the use of ^{125}I-IgG in all immobilization procedures, permits the actual amount of immobilized antigen to be known.

The various immobilization procedures depicted in Figure 3 also permit pre-denatured forms of IgG to be immobilized. Figure 4A illustrates the immobilization on SA-PS of native IgG, IgG denatured for 4 hr at R.T. with 6 M guanidine-HCl (Gu-HCl) or after incubation with Gu-HCl plus 0.5% β-mercaptoethanol. Figure 4A also shows that only biotinylated IgG

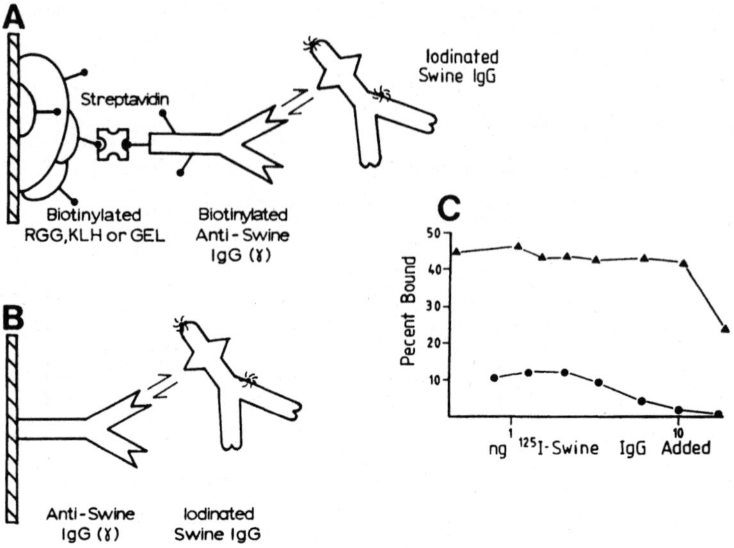

FIGURE 1.

The effect of non-adsorptive immobilization of CAbs on retention of AgCC. Illustrations labelled A and B on the left show how a CAb is either adsorbed directly on polystyrene (B) or immobilized via a streptavidin bridge (A) to some adsorbed, biotinylated carrier protein (RGG, KLY or GEL). Graph C on the right, shows that the AgCC (measured as the percentage of antigen captured) is > 5-fold higher for non-adsorbed CAbs (closed triangles) than for those directly adsorbed. So that the amount of CAb immobilized by each method could be precisely known, initial studies were conducted using ^{125}I-CAb or biotinylated ^{125}I-CAb (6, 29).

becomes immobilized on SA-PS. Figure 4B shows that these differentially-treated IgGs are also adsorbed to polystyrene even in the presence of the denaturing solution.

Our laboratory has investigated the capacity of monoclonal antibodies (mAb) and polyclonal antibodies (pAbs) to recognize various subclasses and allotypes of bovine IgG immobilized as described in Figure 3. Representative data for two mAbs and two pAbs, specific for bovine IgG1, are given in Figure 5. First, these data show that in all cases, IgG1 adsorbed to either Immulon 2®[2] or PEP, is less antigenic than when immobilized on SA-PS. Second, IgG1 pre-denatured with Gu-HCl, is significantly less detectable when immobilized on SA-PS,

---

[2]Immulon 2 is a trademark product of Dynatech, Chantilly, VA. It is a γ-irradiated polystyrene. It has a surface tension component (mJoule/m^2) of -18.3 whereas Polysorp (pure polystyrene) has a component of -0.65 and PEP -0.12. Surface tension components more negative than 28 mJ/m^2 are considered hydrophilic surface and those approaching zero (such as PEP) are regarded as very hydrophobic. The silicone elastomer PEP was supplied by Dow-Corning, Midland, MI, and is of the same chemistry as the envelope of silicone breast implants. Polysorp was obtained from NUNC, Roskilde, Denmark.

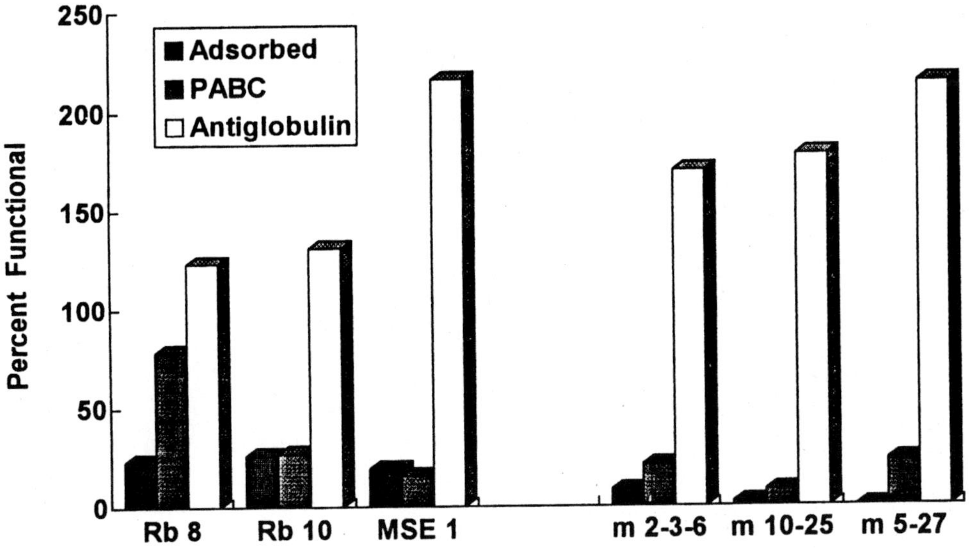

FIGURE 2.

The proportion of three polyclonal (left) and three monoclonal (right) Cabs for fluorescein (FLU) remaining active after being immobilized by different methods. Rb8 = rabbit 8; Rb10 = rabbit 10; Mse1 = mouse 1 and m2-3-6, 10-25 and 5-27 are monoclonal antibodies to FLU. Data are expressed as the proportion of capture antibody equivalents (CAbeqv) which remain active. One CAbeqv corresponds to a bivalent antibody binding one fluorescyl-BSA molecule. CAbeqv approaching 200 are interpreted to mean that each antibody binding site has bound a separate fluorescyl-BSA molecule. PABC = immobilized via the streptavidin bridge configuration illustrated in Figure 1. Antiglobulin immobilization involves the adsorption of a polyclonal anti-mouse or anti-rabbit IgG which then captures the anti-fluorescein antibody. The amount of each CAb immobilized using the different methods was initially determined using ^{125}I-CAbs. Iodinated fluorescyl$_{4.2}$-BSA was used as the antigen. The functional concentration of each anti-FLU was determined by fluorescence quenching. From Butler et al. (29)

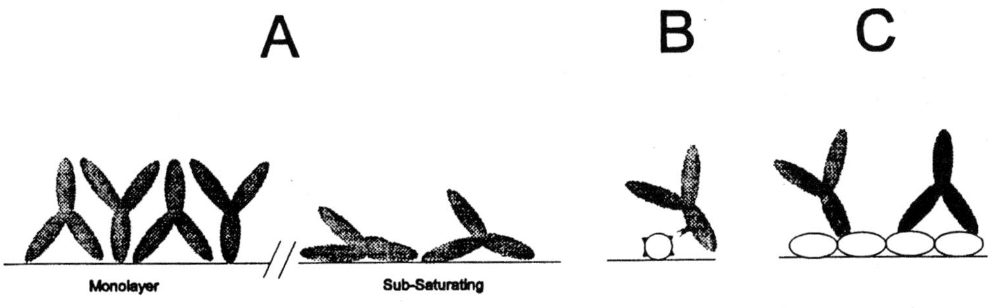

FIGURE 3.

Three different methods for immobilizing bovine IgGs for studies on the antigenicity of immobilized antigens. A= direct adsorption at two different concentrations; B = immobilization via streptavidin; the round, barbed circle depicts streptavidin; C = secondary adsorption; the flattened ellopsoids on the surface depict albumin. The actual amount of each immobilized IgG was determined using ^{125}I-IgG and measuring the radioactivity in each well after completion of the immunoassay. The methods are further described in the text.

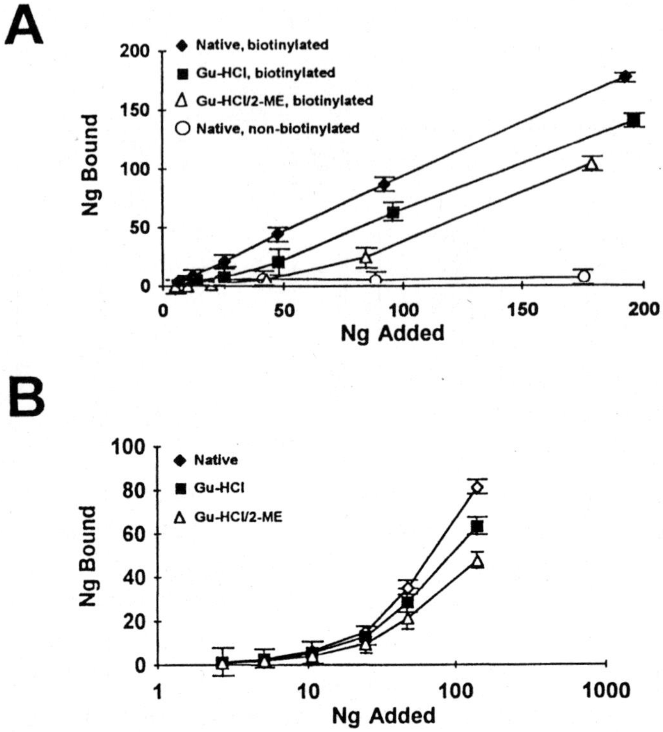

FIGURE 4.

The effect of pre-denaturation of IgGs on their immobilization via a streptavidin bridge (A) or adsorption on polystyrene (B). Data indicate that only biotinylated IgG is immobilized via the streptavidin bridge in the presence of PBS-T and that both native and denatured forms of IgG adsorb to polystyrene, the latter even in the presence of denaturants. Immobilization was determined using ^{125}I-IgGs.

especially when mAbs are used for detection. Third, tnere is no antigenic difference between IgG1 adsorbed in native form or adsorbed after pre-treatment with Gu-HCl. In all cases, whether adsorbed or immobilized on SA-PS, IgG1 which had been reduced with 2-mercaptoethanol in the presence of Gu-HCl, was antigenically inactive (data not shown). Finally, some antibodies (mAb3.7) distinguish between IgG1 adsorbed on polystyrene versus IgG1 adsorbed on the silicone elastomer.

These results show that adsorption alone results in the same degree of antigenic change as pre-denaturation plus adsorption on hydrophobic surfaces. Data also indicate that not all surfaces have the same effect on antigenicity and as might be expected, differences in antigenicity are more easily observed when mAbs are used. Whether these differences in antigenicity are solely due to the greater accessibility of the IgGs immobilized on SA-PS versus when adsorbed, is addressed below.

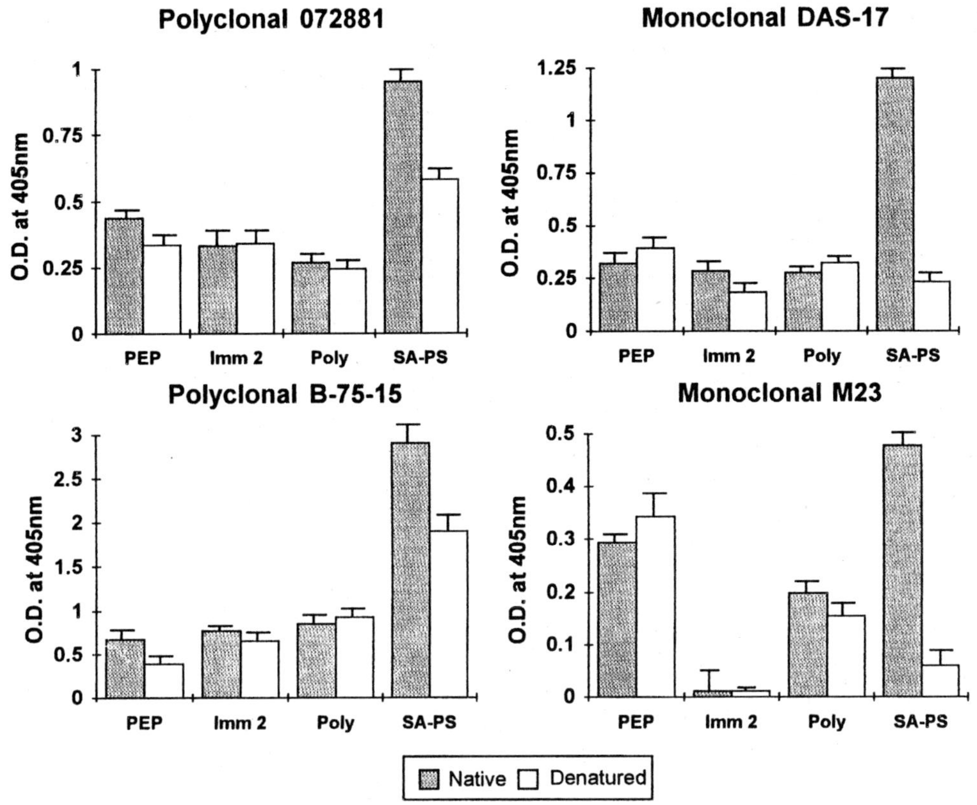

FIGURE 5.

The antigenicity of two polyclonal and two monoclonal antibodies to bovine IgG1 adsorbed to two different types of polystyrene (Imm 2 and Polysorp), a silicone elastomer (PEP) or immobilized via a streptavidin bridge (SA-PS). The histogram values are the mean and standard deviation of the OD405 calculated by linear regression from log-log titration plots, for an amount of immobilized IgG1 (5-20 ng/well) near the midpoint of each log-log titration. Log-log titration plots were linear and parallel in the comparisons studied. Because of differences in the optical properties of the different surfaces, enzymatic reactions were stopped with NaOH and the soluble phase transferred to a common microtiter plate for comparative OD405 measurement. From Butler et al. (8).

Adsorption does not necessarily result in loss of antigenicity. This is demonstrated by the preferential recognition of the A2 allotype of IgG2a(A2) when adsorbed on either polystyrene or the silicone elastomer or the preferential recognition of the A1 allotope when IgG2a is adsorbed on polystyrene (Fig. 6). These results argue against any general concept that epitope accessibility is greater for IgGs immobilized on SA-PS than when adsorbed.

Rescue of antigenicity can also be achieved by allowing the antigen of interest to adsorb to a preformed protein monolayer (Fig. 3C). Figure 7 shows the greater antigenicity of

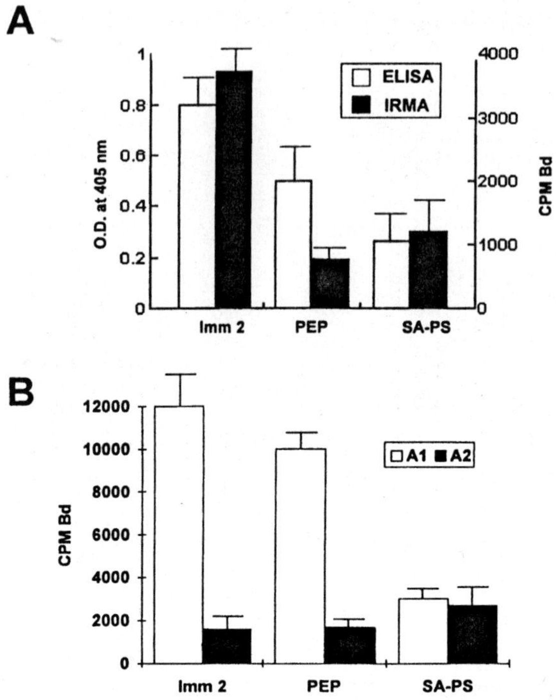

FIGURE 6.

The preferential recognition of allotypic variants of bovine IgG2a when adsorbed on polystyrene (Imm 2) or silicone (PEP) versus when immobilized via a streptavidin bridge (SA-PS). Antigenicity was measured by both ELISA (left ordinate) or using an immunoradiometric assay (IRMA; right ordinate). A = Detection of IgG2a(A2) by monoclonal m574. B = Detection of IgG2a(A1) and IgG2a(A2) by a polyclonal anti-IgG2a. ELISA values expressed were obtained in the same manner as described for the data in Figure 5. When an IRMA was used, the amount of unlabeled IgG adsorbed was calculated from studies using iodinated IgG as described previously (29).

secondarily immobilized IgG versus IgG directly adsorbed to a hydrophobic polymer, especially at very low concentrations of adsorbed IgG.

One argument that can be raised against the data of Figures 5 and 7, is that difference in antigenicity are not a result of conformational change but rather result from: (a) steric hindrance of the large ELISA detection system which has a molecular size of >1000 kd or (b) a chemically unfavorable environment for antibody binding. The latter includes such factors as charge-charge repulsion and physical interference by adjacent molecular structures; the latter is a type of steric hindrance on a delicate scale. The first possibility can be partially tested by employing a detection system equal in size or smaller than the target IgG antigen. Hence, the ^{125}I-fragment of ^{125}I-IgG antibody was used. The Fab fragments were prepared using solid-phase papain (Pierce Chemicals, Rockford, IL) followed by high pressure affinity

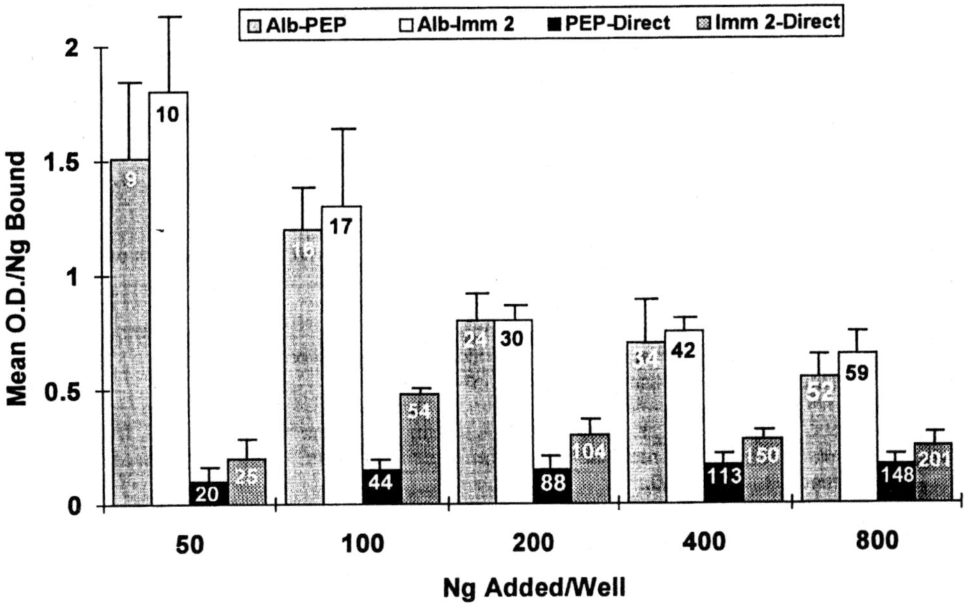

FIGURE 7.

The antigenicity of human IgG primarily or secondarily adsorbed to polystyrene or a silicone elastomer over a range of concentrations. Secondary adsorption was to an albumin monolayer (see Fig. 3C). The legend is on the figure. The values within the bars of the histogram indicate the actual amount of IgG bound. The X-axis indicates the amount of human IgG added either directly to the hydrophobic polymer or to the albumin monolayer. Values on the ordinate were calculated as a ratio of the mean OD405 (see Figure 5 legend) to the ng of IgG immobilized.

chromatography (HiPACA, Chromatochem, Missoula, MT) and subsequent evaluation in denaturing, non-reducing polyacrylamide gels followed by autoradiography (9). Data in Figure 6B indicate that the use of a smaller detection system yielded the same results as those obtained by ELISA. The same was shown for secondarily adsorbed IgG in Figure 7 (9). Hence the differences in antigenicity we report here for adsorbed versus non-adsorbed IgGs, are not artifacts produced by using an ELISA-based detection system which is many-fold larger than the target antibody to which it binds.

Concentration-dependence of Antigenicity of Adsorbed Proteins

Data in Figure 7 suggest that the antigenicity of low concentrations of adsorbed IgG is 5-10 fold lower than equivalent concentrations of IgG that have been secondarily adsorbed on an albumin monolayer. These data also suggest that while secondarily-adsorbed IgG is progressively less antigenic as more is adsorbed, directly adsorbed IgG behaves with "constant antigenicity" (measured as O.D./ng bound) over the range tested. Since the allotopes of IgG2a appear to be better expressed when adsorbed on polystyrene, we wondered what would occur if the antigenicity of adsorbed IgG2a was tested over a range of concentrations with several anti-IgG2a mAbs. Figure 8 compares the antigenicity of IgG2a(A2) adsorbed over the range from 12-456 ng/well using four different anti-IgG mAbs. The concentrations of 12 and 135 are sub-saturating levels (Fig. 3A). While VZ3.12.3d and DAS-2 show no difference in their detection ability over the range studied, m574 especially, but also DAS-9, detect a striking increase in antigenicity when IgG2a(A2) is adsorbed at sub-saturating concentrations.

Preliminary Studies Using Antibody Display Libraries

A popular alternative to the production of conventional mAb is the use of synthetic phage display libraries expressing single chain antibody fragments (scFV; 10). Since it appears that adsorbed proteins may express new epitopes or at least express them in a more favorable manner for recognition (Fig. 6), we believed that phage display libraries could be cloned on adsorbed protein monolayers to obtain clones that recognized only epitopes on adsorbed proteins. These phage clones could then be used to study AICC. Prior to proceeding to full-scale cloning, we tested the ability of an anti-fluorescein (FLU) phage clone provided with the LOX library (11) to recognize FLU-conjugated proteins on various surfaces.

Figure 9 shows that the specificity of this anti-FLU phage was dependent on the compound used for conjugating the fluorescein hapten (FLU) to the adsorbed carrier protein and on the carrier protein used. Figure 9A shows that the anti-FLU phage readily binds to fluorescein isothiocyannate conjugated to BSA (FITC-BSA) but not when the FLU is coupled to BSA using ditriazinoamino fluorescein (DTAF-BSA). However, DTAF-Hgg or BSA conjugated to FITC using a six-carbon spacer (SACF), were also well recognized.

We then wondered if the difference in antigenicity of DTAF-BSA and DTAF-Hgg could be due to the microenvironment of the lysine residues to which FLU should be attached in both proteins. Figure 9B shows that digestion of DTAF-BSA improves its ability to inhibit binding to the solid-phase FITC-BSA versus non-trypsin treatment. However, if BSA is heavily substituted with FLU using DTAF, it becomes a good inhibitor. In the same study, DTAF-Hgg or protease K-treated DTAF-Hgg inhibit equally well. A consistent observation is that fluorescyl-proteins, especially DTAF-Hgg, maintain their inhibitory ability even at high dilution by comparison to the free hapten (FLU-NH_2).

DISCUSSION

The loss of antibody activity by adsorbed anti-FLU CAbs compared to their counterparts immobilized at some distance from the surface, either using a streptavidin bridge (Fig. 1 & 2) or an antiglobulin to immobilize the CAb (Fig. 2), suggests that accessibility might best explain the differences in antigenic activity. This conclusion is also consistent with the lower antigenicity of IgGs adsorbed on PEP or polystyrene as a monolayer versus IgG immobilized on SA-PS or secondarily adsorbed on an albumin monolayer (Figs. 5 & 7). Since lower amounts of IgG can be immobilized on SA-PS versus adsorption as a monolayer (9) one expects the immobilized

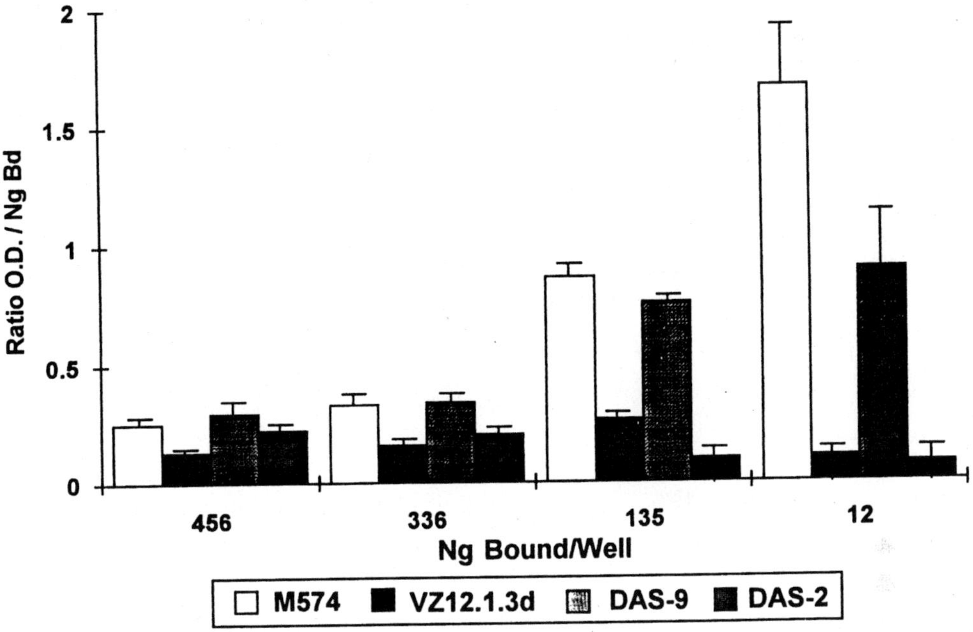

FIGURE 8.

The detectability of bovine IgG2a(A2) by four different monoclonal antibodies when adsorbed to polystyrene at concentrations of 12-456 ng/well. The legend for the four monoclonal antibodies is given on the figure. Values on the ordinate were calculated as described for data presented in Figure 7.

IgG to be packed less tightly on SA-PS than when adsorbed directly. Alternatively, losses in either antibody activity (Figs. 1, 2) or antigenic activity (Fig. 5) could also be a consequence of AICC. Support for this view comes from the data in Fig. 6 in which the A2 and A1 allotopes of IgG2a are more detectable after adsorption than after immobilization on SA-PS. Furthermore, the use of a smaller detection system, i.e. the ^{125}I- Fab of the antibody, did not improve the detectability of adsorbed IgGs. This conclusion would be consistent with reports of physical chemical changes in protein following their adsorption on hydrophobic polymers (12-14) or of mAb which preferentially recognize antigens adsorbed on polystyrene (15-18). However, data in Figure 7 would imply that the degree of AICC remains relatively constant (O.D./ng bound) whereas the detectability of secondarily adsorbed IgGs decreases with increasing concentration. This returns the argument to one which favors the epitope accessibility model. However, when the antigenicity of IgG2a(A2) adsorbed over a range of concentrations was tested with mAbs that recognize IgG2a (Fig. 8), only DAS-9 (anti-λ) and m574 (same mAb used in studies reported in Fig. 6) show a significant increase in detectability when sub-saturating amounts of IgG2a(A2) were adsorbed. Since all mAb did not show this effect, it suggests that

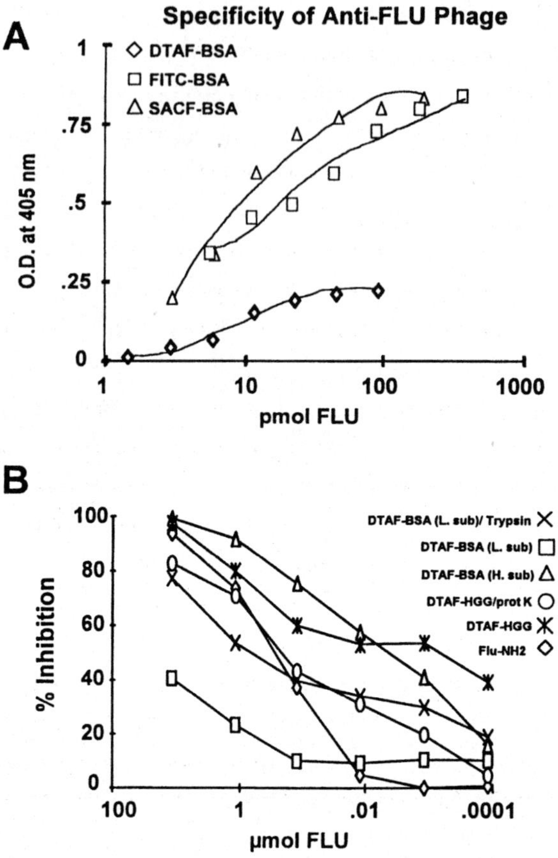

FIGURE 9.

The specificity of an anti-FLU filamentous phage for different fluorescyl protein conjugates. A. Recognition of FITC-BSA, SACF-BSA or DTAF-Hgg but not poorly substituted DTAF-BSA. B. The ability of various fluorescyl conjugates and FLU-NH_2 to inhibit the binding of anti-FLU phage to DTAF-Hgg adsorbed on polystyrene. Legend on figure. L-sub and H-sub indicate low and high FLU substitution respectively. /typsin and /prot k indicate the conjugates were treated with trypsin and protease k, respectively. The various fluorescyl protein conjugates were adsorbed on Immulon 2 (Dynatech, Chantilly, VA), dilutions of the phage clones added in PBS-tween containing 20% Superblock (Pierce, Rockford, IL) and the bound phage subsequently detected after washing using polyclonal anti-M13 conjugated to alkaline phosphate (Pharmacia, Piscataway, NJ).

differences are not merely due to epitope accessibility but rather to epitope exposure or accessibility *resulting from conformational change* at sub-saturating levels. Thus we propose that proteins adsorbed at subsaturating levels assume a different conformation from those in the monolayer (Fig. 3A).

The data presented for the anti-FLU phage demonstrate that changes in either conformation or accessibility need not be dramatic to affect detection significantly. The example of the anti-FLU phage (Fig. 9A) demonstrates how the exact same epitope (FLU) can be more or less available to the anti-FLU phage depending on the form of FLU used for conjugation, the degree of substitution and the type of carrier protein. While the anti-FLU phage doesn't recognize BSA substituted with a low amount of FLU, digestion of this DTAF-BSA substantially increases its ability to inhibit binding to DTAF-Hgg or FITC-BSA adsorbed on polystyrene (Fig. 9B). This result is surprising, since fluorescyl-BSA prepared with DTAF is believed to result in the coupling of FLU to the ε-amino group of lysine in the same manner as when FITC is used (19) and lysine can be expected to occur on the surface of both Hgg and BSA. The studies using the anti-FLU phage are only preliminary, but are suggestive that the microenvironment of an epitope can effect antigenicity.

AICC on polystyrene should be a primary concern in the field on immunodiagnostics. First, in terms of mAb screening procedures, second in regard to the selection of antigens used in an adsorbed form for immunodiagnostic tests and third, because of the considerable amount of *functional* CAb that is lost after adsorption. The allotype data presented in Figures 6 and 8 illustrate the first two concerns. Since >80% of the more than 50 conventionally prepared polyclonal antibodies[3] to bovine IgG2a recognize the immunodominant A1 allotope preferentially (20, 21) and none have been identified which recognized the A2 allotypic variant, the anti-bovine IgG2a reagents currently used in serological and immunochemical tests in veterinary immunodiagnostic are most likely biased in favor of the A1 variants. The findings with traditional reagents also suggest that the sequence of the A2 variant (22) does not result in an allotope or that such an allotope(s) is cryptic. However, A2-specific mAbs can be generated if hybridoma supernatants are deliberately screened against IgG2a(A2) adsorbed on polystyrene at low concentration (Fig. 6). It seems that AICC at subsaturating levels is needed to increase the exposure of a cryptic A2 allotope. Not surprisingly then, we have subsequently shown that the allotypic specificity of anti-bovine IgG2a mAb depends on the method used to immobilize the target antigen (23). This example suggests that the method of immobilizing the target antigen can be crucial in assigning the specificity of mAbs. Monoclonal antibodies for use in immunoassays in which the target antigen is not adsorbed on plastic, e.g. FACS analysis, should be secondarily screened against the target antigen immobilized using a non-adsorptive method, e.g. SA-PS.

AICC is important in immunodiagnostics for a second reason since it means that the critical epitopes on some antigens may not be optimally presented if the antigens are adsorbed on hydrophobic polymers like polystyrene or silicone. This means that serological assays designed to measure the titer of antibodies to a particular tissue or microbial antigen may not provide the best diagnostic test if SPI is used with the adsorbed antigen.

---

[3]Most of the >50 reagents tested were raised by the immunization of goats, rabbits, or guinea pigs with IgG2a emulsified in complete Freund's adjuvant rather than via hybridoma technology and screening on adsorbed IgG2a.

Finally, AICC means that the functional concentration of antibodies adsorbed on a hydrophobic surface is 1-2 logs lower than the same antibody in solution. Not only does this lower the effective performance range of an immunoassay developed using adsorbed CAbs, but it also means that for concentration-dependent antibody mediated events, phenomena observed in solution can not be readily extrapolated to events which occur when adsorbed antibodies are used.

Figure 7 presents yet another practical piece of information relevant to the field of immunodiagnostics. First, it is necessary to emphasize that the secondary adsorption procedure illustrated in Figure 3C and used to generate the data of Fig. 7, has been rigorously tested and we have proven that the IgG secondarily added does not displace the albumin in the monolayer (8; Fig. 3C). This argues that the IgG must be secondary adsorbed onto the albumin monolayer. This being the case, the data of Figure 7 suggest that if stable multilayers of an antigen could be produced by adsorption, the most exposed layer would be most antigenic and probably displayed in most native conformation. This would be consistent with the multilayer passivation theory of Matsuda et al. (24) which holds that as multiple layers of protein become adsorbed to hydrophobic polymers, those exposed are in their most native conformation.

Our data indicate that the silicone elastomer used for the envelope of silicone breast implants (supplied by Dow-Corning) and in many internal prosthesis, behaves much like polystyrene. This observation raises some interesting immunobiological questions. First, it indicates that at least the initial monolayer of host protein adsorbed to such prosthesis has most likely undergone some AICC. However, the much greater antigenicity of secondary layers (Fig. 7) suggests that the degree of AICC is less than for proteins directly adsorbed on the polymer. This is consistent with the Matsuda et al. hypothesis. Since silicone prostheses *in vivo* are exposed to very high protein concentrations, it is quite likely that the exposed protein layer(s) show minimal or no AICC. Even if the immune system is capable of recognizing an altered self protein (see below) the opportunity of encountering them on the surface of an implant would appear to be quite low.

Let us now assume that antigenically altered host proteins can indeed be presented to the immune system, e.g. silicone gel from ruptured implants may adsorb protein and then be ingested by APCs such as macrophages. When this occurs, are the peptides of altered host proteins presented by the MHC different from those of native host proteins for which the host immune system is believed to be tolerant? There are several theories suggesting that the peptides of denatured proteins presented by APCs may be different than those presented for native proteins (25). Furthermore, some self-reactive clones are not deleted, but persist or are inactivated and are capable of being activated under appropriate stimulation (26, 27). It seems possible that AICC of host proteins on medical prostheses could result in immunological complications. There is a need to address the various questions which our observations raise in the context of immunological tolerance to self-proteins.

The preliminary observations made using the anti-FLU phage suggest that differences seen in the antigenicity of adsorbed proteins may not result from conformational changes in the epitope itself, but rather from changes in the chemical microenvironment of the epitope. This was also observed by Mummert and Voss (28) using FITC conjugated to lysine on five different protein carriers. This probably also explains the shallow inhibition curves observed using fluorescyl protein compared to $FLU\text{-}NH_2$. Thus, synthetic phage libraries may prove valuable in identifying subtle changes in epitopes or their environment.

The data we have presented suggest that the antigenic differences between adsorbed IgG and IgG non-adsorbed but nevertheless immobilized on a solid-phase, must in part be due to AICC. Thus, AICC is not only relevant in the field of immunodiagnostic but also to surgically-induced immunopathology. Unfortunately, distinguishing between antigenic differences due to the molecular environment surrounding an epitope and conformational changes in the epitope itself, is difficult. We hope that synthetic antibody phage display libraries will aid in resolving this distinction. Like most scientific investigations, we have answered a few questions, but simultaneously raised many more.

ACKNOWLEDGMENT

The research described was supported by a grant from Abbott Laboratories, North Chicago, IL and from the Dow-Corning Corporation of Midland, MI. The authors thank Ms. Marcia Reeve for preparation of the typescript and Professor Carel van Oss for determination of the surface tension energy components of the three adsorptive surfaces used in the study.

ABBREVIATIONS

AICC = adsorption-induced conformational change
APC = antigen presenting cell
BSA = bovine serum albumin
CAb = capture antibody
CAbeqv = capture antibody equivalent; a functional activity equivalent to a CAb with one binding site
DTAF = ditriazinoaminofluorescein
ELISA = enzyme-linked immunosorbent assay
FACS = fluorescence activated cell sorter
FITC = fluorescein isothiocyanate
FLU = fluorescein hapten
Gu-HCl = quanidine hydrochloride
Hgg = human gamma globulin
IRMA = immunoradioactive assay; non-competitive detection using ^{125}I-antibody
mAb = monoclonal antibody
pAb = polyclonal antibody
PEP = platinum catalyzed elastomer of polydimethylsiloxane
RIA = radioimmunoassay; implies classical competitive configuration
SPI = solid-phase immunoassay
SA-PS = streptavidin functionalized polystyrene
SACF = succinimidyl amino caproic fluorescein

REFERENCES

1. Yalow, R. W., and S. A. Berson. J Clin Invest 39, 1157-1168 (1960).
2. Catt, K., and G. W. Tregear. 1967. Science 158, 1570-1572 (1967).
3. Mason, D. W., and A. F. Williams. Biochem J 187, 1-20 (1980).
4. Stenberg, M., and H. Nygren. Anal Biochem 127, 183-192 (1982).
5. Butler, J. E. in Structure of Antigens (M.H.V. van Regenmortel, ed.). CRC Press, Boca Raton, FL. (1991) pp. 209-259.
6. Suter, M., and J. E. Butler. Immunol Lett 13, 313-317 (1986).
7. Cantarero, L. A., J. E. Butler and J. W. Osborne. Anal Biochem 105, 375-383 (1980).
8. Butler, J. E., E. P. Lü, B. Christiansen and P. Navarro. J Mol Recog (1996). (pending)
9. Butler, J. E., P. Navarro and E. P. Lü. J Mol Recog (1996). (pending)
10. Winter, G., and C. Milstein. Nature 349, 293-299 (1991).
11. Griffiths, A. D., S. C. Williams, O. Hartley, I. M. Tomlinson, et al. EMBO J. 13, 3245-3260 (1994).
12. Bull, H. B. Biochim Biophys Acta 19, 464-471 (1956).
13. Burghardt, T. P., and D. Axelrod. Biochemistry 22, 979-985 (1983).
14. Oreskes, I., and J. M. Singer. J Immunol 86, 338-344 (1961).
15. Kennels, S. J Immunol Meth 55, 1-12 (1982).
16. Mierendorf, R. C., Jr., and R. L. Dimond. Analyt Biochem 135, 221-229 (1983).
17. Holland, Z., and E. Katchalski-Katzir. Mol. Immunol. 23, 927-934 (1986).
18. Darst, S. A., C. R. Robertson and J. A. Berzofsky. J Biophysiol Soc 53, 533-539 (1988).
19. Blakeslee, D., and M. G. Baines. J Immunol Meth, 13, 305-320 (1976).
20. Butler, J. E., and H. Heyermann. Mol Immunol 23, 291-296 (1986).
21. Butler, J. E., P. Navarro and H. Heyermann. Vet Immunol Immunopath 40, 119-133 (1994).
22. Kacskovics, I., and J. E. Butler. Molec. Immunol. 33, 189-195 (1996).
23. Navarro, P., E. P. Lü, S. Yahiro, K. L. Nielsen and J. E. Butler. (pending)
24. Matusda, T., H. Takano, K. Hayashi, Y. Taenaka, S. Takaichi, M. Umezu, T. Nakamura, H. Iwata, T. Nakatani, T. Tanaka, S. Talatani and T. Akutsu. Tran Am Soc Artif Intern Org 30, 353-358 (1984).
25. Sercarz, E.E., P. V. Lehmann, A. Ametani, G. Beniehou, A. Miller and K. Moudgil. Ann Rev Immunol 11, 729-766 (1993).
26. Goodnow, C. C. Ann Rev Immunol 10, 489-518 (1992).
27. Lou, Y., and K.S.K. Tung. J Immunol 151, 5790-5799 (1993).
28. Mummert, M. E., and E. W. Voss, Jr. Molec Immunol 32, 1225-1233 (1995).
29. Butler, J. E., L. Ni., W. R. Brown, K. S. Joshi, J. Chang, B. Rosenberg and E. W. Voss, Jr. Mol Immunol 30, 1165-1175 (1993).

Rapid Diagnosis of Periodontal Infections: Findings in AIDS Patients

Joseph J. Zambon, D.D.S., Ph.D.
Professor of Periodontology and Oral Biology
State University of New York at Buffalo
School of Dental Medicine
Buffalo, New York 14214

Abstract

A small number of bacterial pathogens in the human oral cavity cause the different forms of periodontal disease. Of the approximately two hundred different oral bacterial species, about a dozen have been associated with these diseases including localized juvenile periodontitis, rapidly progressing periodontitis, and adult periodontitis. These species include Actinobacillus actinomycetemcomitans, Bacteroides forsythus, Campylobacter rectus, Porphyromonas gingivalis, and Prevotella intermedia. Several rapid methods have been developed to detect these species in clinical samples. These include immunologic methods such as immunofluorescence, nucleic acid assays such as DNA-DNA hybridization in dot blots, and enzyme assays. Immunofluorescence microscopy has been used to determine the prevalence and relative proportions of these pathogens in dental plaque samples from 194 subjects including HIV-infected and uninfected male homosexuals and intravenous drug users.

Introduction

The human oral cavity carries a wide variety of microorganisms including bacteria, viruses, and fungi. Best estimates suggest that 200 to 300 distinct bacterial species can inhabit the mouth (1). The bacteria, in particular, are important microorganisms causing both dental caries and the various forms of periodontal disease. Together these constitute the major oral diseases in Western populations.

Periodontal disease is not a single entity but, rather, a group of diseases which affect the gingiva, connective tissue attachment and alveolar bone. There are, for example, forms of periodontal disease which affect only the gingiva such as acute necrotizing Some types of periodontal disease occur in particular age groups. Disease in adolescents is known as localized and generalized juvenile periodontitis while there are forms which affect adults such as chronic adult periodontitis.

There is general agreement that most periodontal diseases are caused by bacteria in dental plaque (2, 3). While previous concepts proposed that gross accumulations of dental plaque caused periodontal disease - the "non-specific plaque" hypothesis, current concepts propose that only certain bacterial species in dental plaques cause severe forms of the disease. This "specific plaque" hypothesis (4) proposes that dental plaques isolated from periodontitis lesions are qualitatively distinct from those isolated from healthy sites. Therefore, while gingival inflammation can be caused by dental plaque, per se, the loss of connective tissue attachment and alveolar bone which are the hallmarks of periodontitis are associated with only a few oral bacteria. The specific plaque hypothesis is supported by numerous studies examining the composition of subgingival dental plaque from different forms of periodontal disease (4, 5, 6, 7, 8, 9, 10, 11, 12, 13, 14).

Periodontopathic Bacteria

Which of the many oral bacteria are important in causing periodontitis? While Koch's postulates can delineate the etiology of monoinfections, they are less useful in mixed microbial infections such as periodontal disease. Alternate criteria have been developed to identify specific microorganisms in the etiology of human periodontal disease including (15); 1) the presence of high numbers of the microorganism in the lesion compared to either its absence or its presence in low numbers in other sites; 2) elevated levels of serum, salivary, or gingival crevicular fluid antibody to the bacterium; 3) bacterial virulence factors which can be correlated with clinical histopathology; 4) animal models demonstrating tissue destruction in the presence of the microorganism; and, 5) clinical improvement upon elimination of microorganism from periodontal lesions.

Prime among the approximately dozen bacterial species which have been implicated as periodontal pathogens is Actinobacillus actinomycetemcomitans. The association between A. actinomycetemcomitans and localized juvenile periodontitis is such that this microorganism is considered etiologic in most cases (1). A. actinomycetemcomitans is present in high numbers in juvenile periodontitis lesions where it can invade gingival connective tissue (16). By contrast, healthy sites in these same patients harbor little or no A. actinomycetemcomitans. Infected patients demonstrate high levels of antibody against A. actinomycetemcomitans and the bacterium produces several virulence factors including a potent leukotoxin which can destroy host polymorphonuclear leukocytes. Patients improve when A. actinomycetemcomitans is eliminated and their disease either progresses or does not resolve when the microorganism persists in periodontal pockets. There is also some animal model data indicating that A. actinomycetemcomitans can destroy alveolar bone. Altogether, there is compelling evidence that A. actinomycetemcomitans can cause juvenile periodontitis.

There is also compelling evidence that Porphyromonas gingivalis can cause adult periodontitis. High numbers of P. gingivalis can frequently be found in adult periodontitis lesions where it may invade epithelium (17). P. gingivalis exhibits virulence factors including fimbrial adhesins, LPS, capsule, collagenase, and trypsin-like enzymes.

While A. actinomycetemcomitans and P. gingivalis are considered key periodontal pathogens, several other species have been associated with periodontal disease. These species include Bacteroides forsythus, Campylobacter rectus, Eubacterium species, Eikenella corrodens, Fusobacterium nucleatum, Peptostreptococcus micros, Porphyromonas gingivalis, Prevotella intermedia, Selenomonas species and certain oral spirochetes (1, 10, 11, 18, 19, 20).

The initial event in periodontal disease (Figure 1) is the transmission of periodontal pathogens, usually from other family members, to an individual's oral cavity where these microorganisms colonize the gingiva. They persist in gingival sites by evading host defense mechanisms and produce virulence factors which cause alveolar bone loss. Periodontitis is clinically detectable at this point and periodontal treatment can be instituted to both eliminate the pathogenic bacteria and restore the periodontium. However, these infections often recur due to incomplete elimination during therapy or by re-infection.

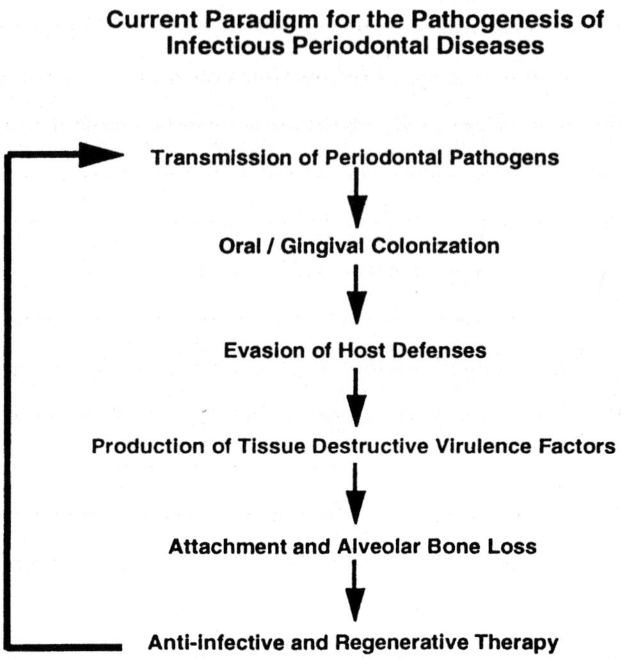

Figure 1

Laboratory Assays for Periodontal Infections

Microscopy, particularly darkfield and phase contrast microscopy, has been used for several decades to determine the size, shape and motility of the individual bacteria in patient plaque samples (21, 22). "Healthy" plaques have small numbers of mainly non-motile cocci and rods while "disease" plaques have large numbers of spirochetes, fusiforms, filaments, and motile forms. Clinical efficacy can be demonstrated by shifts from disease plaques to healthy plaques. However neither darkfield nor phase contrast microscopy can discriminate individual bacterial species. Therefore, it falls short as a method for identifying periodontal pathogens.

Bacterial culture enumerates the broadest spectrum of bacterial species present in patient plaques and is the method of choice for antibiotic susceptibility and resistance. This method if enhanced by the availability of a number of selective media targeted to individual species, for example, TSBV media favors the recovery of A. actinomycetemcomitans.

Although refined by the use of semi-automated methods for distributing samples and better identification methods, culture is cumbersome, time consuming and costly.

Immunologic reagents or nucleic acid probes can be employed in a variety of test configurations in the rapid diagnosis of periodontal infections using species-specific polyclonal antisera or a monoclonal antibody in the case of immunoassays (23), or whole genomic, cloned, or synthetic oligonucleotides in the case of nucleic acid probes (24, 25, 26, 27, 28, 29, 30). Immunoassays for periodontal pathogens include immunofluorescence microscopy (31, 32), membrane immunoassay (33), latex agglutination assays (34), and particle concentration fluorescence immunoassay (PCFIA) which utilizes polystyrene beads as a substrate (35). Synthetic oligonucleotide primers can also be used in the polymerase chain reaction (PCR) to detect as few as 10 A.actinomycetemcomitans cells in a patient plaque sample (36, 37). A variant of PCR known as the arbitrarily primed polymerase chain reaction (AP-PCR) has been used to trace the source of periodontal infections (38, 39). Finally, enzyme assays can be used to detect periodontal pathogens including assays for collagenase, peptidases, trypsin-like enzymes, neutral proteases, and elastase which detect one or more of a group of periodontal pathogens (40, 41).

A major technical roadblock to the identification of additional periodontal pathogens has been the inability to culture a significant proportion of bacteria in dental plaque. Methods in molecular biology have been developed to identify previously non-cultivable bacteria in dental plaque samples. "Universal" primers for prokaryotic 16S DNA can be used in PCR to amplify the intervening nucleic acid sequence from uncultivable plaque bacteria. The 16S ribosomal RNA sequences resulting from the PCR can be analyzed, and identified by comparison to a database of other 16S RNA sequences.

Periodontal pathogens can be detected with increasing prevalence in deep periodontal pockets. Wolff et al. (42) found that in deep periodontal pockets the odds of finding P. gingivalis was 3.9, A. actinomycetemcomitans - 3.0, P. intermedia - 4.0, E. corrodens - 2.7, and F. nucleatum - 2.8 . The prevalence of P. gingivalis, P. intermedia and A. actinomycetemcomitans increases with pocket depth and P. gingivalis is elevated in suppurating sites (43). Elevated subgingival temperature is associated with P. intermedia, A. actinomycetemcomitans, and P. gingivalis (44).

Assays for periodontal pathogens are not, themselves, diagnostic for periodontal disease. Periodontal bacterial pathogens can be present in high numbers in periodontal pockets without loss of connective tissue attachment or alveolar bone. Subgingival infection can increase the risk for periodontitis. For example, subgingival infection with P. gingivalis and B. forsythus increase the relative risk for periodontitis by 1.59 and 2.45 fold, respectively (45).

Rapid Detection of Periodontal Pathogens in AIDS Patients

An example of the utility of rapid methods for the diagnosis of periodontal infections can be seen in studies of the periodontal lesions in subjects infected with Human Immunodeficiency Virus (HIV). The etiology of the periodontal lesions associated with HIV infection remain obscure. These lesions include an atypical gingivitis known as linear gingival erythema (formerly called HIV-gingivitis) and a particularly rapid and severe form of periodontitis known as necrotizing periodontitis (formerly called HIV-periodontitis). That these lesions are, in fact, of microbial origin can be derived from the effect of antimicrobial therapy in ameliorating the clinical course. Linear gingival erythema which is thought to represent a form of gingival candidiasis responds to antifungal agents while necrotizing periodontitis responds to antibiotics including metronidazole.

What do we know about the bacteria associated with these diseases? Two points emerge from the literature. First, HIV-infected individuals harbor the same periodontal pathogens as are found in non-HIV-infected individuals including Actinobacillus actinomycetemcomitans, Bacteroides forsythus, Campylobacter rectus, Eubacterium species, Porphyromonas gingivalis, and Prevotella intermedia . Second, HIV-infected individuals also harbor other microorganisms some of which are pathogenic and which are not usually found in HIV-uninfected individuals. For example, HIV-infected individuals can have increased levels of Mycoplasma salivarium and yeast in their oral cavities. They can also have enteric species such as Enterobacter cloacae and Enterococcus faecalis as well as clostridia including Clostridium difficile and as yet undefined groups of clostridia which have been found exclusively in injecting drug users.

To further characterize the oral and periodontal bacteria in AIDS patients, we conducted the following study. Plaque samples were taken from the mesial surface of the four permanent first molar teeth by means of a sterile paper point then placed into 1 ml of sterile Ringer's solution with 2% formalin. Ten microliter aliquots of the dispersed bacterial suspension were distributed onto glass slides, heat-fixed and reacted with working titer concentrations (the highest two-fold serial dilution still giving 4+ fluorescence) of species-specific rabbit antisera to Actinobacillus actinomycetemcomitans, Actinomyces viscosus, Prevotella intermedia, Bacteroides forsythus, Candida albicans, Eikenella corrodens, Eubacterium saburreum, Fusobacterium nucleatum, Porphyromonas gingivalis, Streptococcus mutans, Streptococcus sanguis, and Campylobacter rectus. The bacterial smears were then allowed to react with affinity purified goat anti-rabbit IgG conjugated to fluorescein isothiocyanate (FITC isomer I, fluorescein to protein ratio of 25 μg/ml, BBL Microbiology Systems, Cockeysville, MD). The slides were mounted with glycerol in PBS (2:1 v/v) and examined with a Zeis standard 14 microscope equipped for phase contrast illumination and for incident light fluorescence. Fluorescence was graded from 0 to 4+, with grades 3+ and 4+ considered serologically positive reactions. Between 100 and 200 bacterial cells were counted on each sample. Plaque smears were considered positive for a microorganism if they exhibit fluorescent microorganisms with well-defined cell outlines of a size and shape consistent with the test microorganism and with dark or lightly fluorescing centers, and which constitute more than 1% of the total cell count determined by phase contrast microscopy. Positive controls included bacterial smears of pure cultures stained with the respective species-specific antisera and fluorescein-conjugated antisera.

Tables 1 and 2 show the results of immunofluorescence analysis of dental plaque samples from 64 HIV-infected and 45 HIV-uninfected homosexual men and from 49 HIV-infected and 36 HIV-uninfected intravenous drug users. In the HIV-infected homosexual men, A. viscosus was the most prevalent of the target species being found in all of these subjects and it also constituted the highest mean proportion of the bacteria at 2.44%. S. sanguis was the next most prevalent of the target species, 97%, followed by P. intermedia, 95%, Capnocytophaga species, 88%, C. rectus and S. mutans at 73%, and B. forsythus, 72%. S. sanguis comprised the next highest mean proportion, 1.11%, followed by P. intermedia, 0.72%, S. mutans, 0.59%, Capnocytophaga species, 0.44%, and B. forsythus, 0.33%.

In the HIV-uninfected homosexual men, A. viscosus was again the most prevalent of the target species, 100%, followed by P. intermedia, 98%, S. sanguis, 96%, Capnocytophaga species, 93%, C. rectus, 84%, S. mutans, 84%, and B. forsythus 78%. A. viscosus also comprised the

TABLE I

HIV-Infected Homosexual Men (N=64)

Target Species	Percent Positive Subjects	Level of Infection (%) Mean	Range		
A. actinomycetemcomitans	39	0.06	0.00	-	0.64
A. viscosus	100	2.44	0.49	-	7.34
B. forsythus	72	0.33	0.00	-	1.78
P. gingivalis	67	0.40	0.00	-	2.14
P. intermedia	95	0.72	0.00	-	3.37
Capnocytophaga species	88	0.44	0.00	-	1.78
E. corrodens	48	0.10	0.00	-	1.28
E. saburreum	30	0.14	0.00	-	0.55
F. nucleatum	41	0.12	0.00	-	1.60
S. mutans	73	0.59	0.00	-	4.04
S. sanguis	97	1.11	0.00	-	6.87
C. rectus	73	0.22	0.00	-	2.17

HIV Uninfected Homosexual Men (N=45)

Target Species	Percent Positive Subjects	Level of Infection (%) Mean	Range		
A. actinomycetemcomitans	44	0.06	0.00	-	0.64
A. viscosus	100	2.03	0.42	-	5.02
B. forsythus	78	0.49	0.00	-	2.37
P. gingivalis	62	0.39	0.00	-	2.47
P. intermedia	98	1.00	0.00	-	6.68
Capnocytophaga species	93	0.49	0.00	-	1.87
E. corrodens	56	0.14	0.00	-	1.64
E. saburreum	67	0.14	0.00	-	0.49
F. nucleatum	51	0.17	0.00	-	1.02
S. mutans	84	0.33	0.00	-	3.31
S. sanguis	96	0.91	0.00	-	3.14
C. rectus	84	0.29	0.00	-	1.35

highest mean proportion of the total bacterial cell count, 2.03%, followed by P. intermedia, 1%, and S. sanguis, 0.91%.

In the HIV-infected intravenous drug users, both A. viscosus and P. intermedia were found in all of these subjects. Next most prevalent were B. forsythus, 94%, S. sanguis, 92%, S. mutans, 92%, Capnocytophaga species, 84%, C. rectus, 82%, and P. gingivalis, 80%. A. viscosus constituted the highest mean proportion of the bacteria in these samples, 1.88, followed by P. intermedia, 1.7%, B. forsythus, 1.03%, and P. gingivalis, 0.76%.

In the HIV-uninfected intravenous drug users, A. viscosus was found in highest prevalence at 100% followed by P. intermedia at 97%, B. forsythus and S. mutans both at 94%, E.

TABLE II

HIV Infected IV Drug Users (N=49)

Target Species	Percent Positive Subjects	Level of Infection (%) Mean	Range
A. actinomycetemcomitans	39	0.07	0.00 - 1.00
A. viscosus	100	1.88	0.14 - 4.14
B. forsythus	94	1.03	0.00 - 2.44
P. gingivalis	80	0.76	0.00 - 5.34
P. intermedia	100	1.70	0.36 - 5.99
Capnocytophaga species	84	0.26	0.00 - 1.12
E. corrodens	65	0.10	0.00 - 1.31
E. saburreum	94	0.26	0.00 - 0.95
F. nucleatum	63	0.32	0.00 - 2.56
S. mutans	90	0.65	0.00 - 3.14
S. sanguis	92	0.77	0.00 - 4.02
C. rectus	82	0.40	0.00 - 1.89

HIV Uninfected IV Drug Users (N=36)

Target Species	Percent Positive Subjects	Level of Infection (%) Mean	Range
A. actinomycetemcomitans	47	0.12	0.00 - 1.58
A. viscosus	100	1.61	0.45 - 5.38
B. forsythus	94	1.08	0.00 - 2.62
P. gingivalis	83	0.87	0.00 - 6.76
P. intermedia	97	1.29	0.00 - 4.82
Capnocytophaga species	72	0.15	0.00 - 0.67
E. corrodens	44	0.10	0.00 - 1.06
E. saburreum	86	0.19	0.00 - 1.08
F. nucleatum	44	0.12	0.00 - 1.03
S. mutans	94	0.74	0.00 - 4.85
S. sanguis	89	0.46	0.00 - 1.73
C. rectus	78	0.28	0.00 - 2.64

saburreum at 86%, and P. gingivalis at 83%. A. viscosus constituted the highest mean proportion of the bacteria in these samples, 1.61%, followed by P. intermedia, 1.29%, B. forsythus, 1.08%, and P. gingivalis, 0.87%.

Summary

Rapid microbiological assays have spurred the application of the infectious disease paradigm to diagnosis and treatment of periodontal disease. Continued progress in the area of periodontal microbiology and in the laboratory diagnosis of periodontal infections will further our understanding of the complex microbial ecology which exists subgingivally, and help define the host-parasite interactions associated with active periodontal disease.

References

1. Haffajee AD, Socransky SS. Periodontology 2000, vol 5, 78-111 (1994)

2. Socransky SS, Haffajee AD. J Periodontol, 63, 322-331 (1992)

3. Dzink JL, Socransky SS, Haffajee AD. J Clin Periodontol, 15, 316-323(1988)

4. Loesche WJ. Chemotherapy of dental plaque infections. Oral Sci Rev ,9, 65-107, (1976)

5. Gusberti FA, Mombelli A, Lang NP, Minder CE. J Clin Periodontol ,17, 685-692 (1990)

6. Listgarten MA, Lai CH, Young V. J Periodontol, 64, 155-161(1993)

7. Loesche WJ. Adv Dent Res, 2, 275-283 (1988)

8. Slots J. Importance of Black-Pigmented Bacteroides in Human Periodontal Disease. Washington, D.C, Am Soc Microbiol (1982)

9. Tanner A, Bouldin H. J Clin Periodontol, 15, 467-471(1989)

10. van der Weijden GA, Timmerman MF, Reijerse E, Wolffe GN, van Winkelhoff AJ, van der Velden U. J Clin Periodontol, 21, 583-588(1994)

11. van Winkelhoff AJ, van Steenbergen TJM, de Graaff J. J Clin Periodontol, 15, 145-155 (1988)

12. Zambon JJ. J Clin Periodontol, 12, 1-20 (1985)

13. Zambon JJ, Reynolds H, Fisher JG, Shlossman M, Dunford R, Genco RJ. J Periodontol, 59, 23-31 (1988)

14. Zambon JJ, Sunday GJ, Smutko JS. J Periodontol, 61, 75-80.(1990)

15. Socransky SS. J Periodontol, 48,497-504 (1977).

16. Saglie FR, Marfany A, Camargo P. J Periodontol, 59, 259-265 (1988)

17. Sandros J, Papapanou PN, Nanmark U, Dahlen G. J Periodontal Res, 29, 62-69 (1994)

18. Mombelli A, Gmur R, Gobbi C, Lang NP. J Periodontol, 65, 820-826 (1994)

19. Dahlen G, Manji F, Baelum V, Fejerskov O. J Clin Periodontol, 16, 305-310 (1989)

20. Simonson LG, Robinson PJ, Pranger. J Periodontol, 63, 270-273 (1992).

21. Omar AA, Newman HN, Bulman J, Osborn J. J Clin Periodontol, 18, 555-566 (1991).

22. Lopez NJ, Mellado JC, Giglio MS. Leighton GX. J Periodontol, 66, 559-567 (1995)

23. Kamiya I, Okuda K, Hara K. J Periodontol, 65, 309-315 (1994)

24. Kojima T. Kokubyo Gakkai Zasshi , 61, 71-81 (1994)

25. Dix K, Watanabe SM, McArdle S, Lee DI, Randolph C, Moncla B, Schwartz DE. J Clin Microbiol , 28, 319-323 (1990)

26. Loesche WJ, Giordano JR, Hujoel P, Schwarcz J, Smith BA. J Clin Periodontol, 19, 103-112 (1992)

27. Melvin WL, Assad DA, Miller GA, Gher ME, Simonson L, York AK. J Periodontol , 65, 576-582 (1994)

28. Bolstad AI, Skaug N, Jensen HB. J Periodont Res , 26, 519-526 (1991)

29. Haffajee AD, Socransky SS, Smith C, Dibart S. J Clin Periodontol , 19, 84-91 (1992)

30. Lotufo RF, Flynn J, Chen C, Slots J. Oral Micro & Immunol , 9, 154-160 (1994)

31. Bonta Y, Zambon JJ, Neiders M, Genco RJ. J Dent Res , 64, 793-798 (1985)

32. Slots J, Hafstrom C, Rosling B, Dahlen G. J Periodont Res, 20, 613-625 (1985)

33. Snyder B, Zambon JJ, Reynolds HS, Ryerson CC, Genco RJ. J Dent Res, 73, 305.(1994)

34. Nisengard RJ, Mikulski L, McDuffie DM, Bronson, P. J Periodontol , 63, 611-617 (1992)

35. Wolff LF, Anderson L, Sandberg, GP, Reither L, Binsfield CA, Corinaldesi G, Shelburne CE. J Periodontol , 63, 1093-1101 (1992)

36. Lin CY, Wong MY, Jeng JH, Chang WK, Kuo MY. J Formosan Med Assoc, 93, 289-293 (1994)

37. Leys EJ, Griffen AL, Strong. J Clin Microbiol, 32, 1288-1294 (1994)

38. Preus HR, Haraszthy VI, Zambon JJ, Genco RJ. J Clin Microbiol, 31, 2773-2776 (1993)

39. Preus HR, Zambon JJ, Dunford RG, Genco RJ. J Periodontol , 65, 2-7 (1994)

40. Loesche WJ. J Periodontol, 63, 1102-1109 (1992)

41. Pederson ED, Miller JW, Matheson S, Simonson LG, Chadwick DE, Covill PJ, Turner DW, Lamberts BL, Morton HE. J Clin Periodontol, 21, 519-525 (1994)

42. Wolff LF, Aeppli DM, Pihlstrom B, Anderson L, Stoltenberg J. J Clin Periodontol, 20, 699-706 (1993)

43. Socransky SS, Haffajee AD. J Periodontal Res, 26, 195-212 (1991)

44. Haffajee AD, Socransky SS, Smith C, Dibart S, Goodson JM. J Clin Periodontol, 19, 417-422 (1992)

45. Grossi SG, Zambon JJ, Ho AW, Koch G, Dunford RG, Machtei EE, Norderyd OM, Genco RJ. J Periodontol, 65, 260-267 (1994)

USES OF BIOSENSORS IN THE STUDY OF VIRAL ANTIGENS

Marc H.V. Van Regenmortel, Danièle Altschuh, Jean Chatellier,
Nathalie Rauffer-Bruyère, Pascale Richalet-Sécordel and Hélène Saunal
Immunochemistry Department
CNRS, Institut de Biologie Moléculaire et Cellulaire
15 rue René Descartes, 67084 Strasbourg Cedex, France

ABSTRACT

The introduction in 1990 of a new biosensor technology based on surface plasmon resonance has greatly simplified the measurement of binding interactions in biology. This new technology known as biomolecular interaction analysis makes it possible to visualize the binding process as a function of time by following the increase in refractive index that occurs when one of the interacting partners binds to its ligand immobilized on the surface of a sensor chip. None of the reactants needs to be labelled, which avoids the artefactual changes in binding properties that often result when the molecules are labelled. Biosensor instruments are well-suited for the rapid mapping of viral epitopes and for identifying which combinations of capturing and detector Mabs will give the best results in sandwich assays. Biosensor binding data are also useful for selecting peptides to be used in diagnostic solid-phase immunoassays. Very small changes in binding affinity can be measured with considerable precision which is a prerequisite for analyzing the functional effect and thermodynamic implications of limited structural changes in interacting molecules. On-rate (k_a) and off-rate (k_d) kinetic constants of the interaction between virus and antibody can be readily measured and the equilibrium affinity constant K can be calculated from the ratio $k_a/k_d = K$.

INTRODUCTION

The introduction in 1990 of a new biosensor technology based on surface plasmon resonance (SPR) has greatly simplified the measurement of binding interactions in biology. This new technology known as biomolecular interaction analysis (BIA) makes it possible to visualize the binding process as a function of time by following the increase in refractive index that occurs when one of the interacting partners binds to its ligand immobilized on the surface of a sensor chip. None of the reactants needs to be labelled, which avoids the artefactual changes in binding properties that often result when the molecules are labelled. Several reviews describing this new technology are available (1-3) and special issues of journals have been devoted to the many applications of biosensors in biochemistry and immunology (4, 5). In this review, we will describe how this technology can be used for studying viral antigens and for developing probes intended for viral diagnosis.

The first commercialized biosensor instrument based on SPR technology was the BIAcore™ developed by Pharmacia Biosensor (Uppsala, Sweden). An upgraded version, the BIAcore 2000, was released in 1995 and a simpler, less automated version of the instrument, the BIAcore X, became available in 1996. Another instrument based on a resonant mirror system (6) and called IAsys is commercialized by Affinity Sensors (previously Fisons Applied Sensor Technology, Cambridge, U.K.). A comparison between some of the available systems has been published (7).

Compared to other immunoassay systems, biosensor instruments have the following advantages:

1) The immobilized ligand is covalently bound to a hydrophilic dextran matrix present on the sensor chip which tends to preserve the conformation of the molecule. In contrast, in classical solid-phase immunoassays, the ligand is immobilized by adsorption to plastic and this usually induces at least partial denaturation of proteins and a concomitant alteration of epitopes (8).
2) Since none of the reactants needs to be labelled, the alterations in biological activity that often result from the conjugation process are avoided.
3) Every step in a multiple-layer assay can be directly visualized and it is immediately apparent if one of the components of the system is defective. In contrast, in assays that use labelled antibodies, only the last step is detectable and no direct information is provided on the course of earlier binding steps in the reaction sequence. If the end result is unsatisfactory, each component of an ELISA system must be analyzed separately to discover which one is at fault. When the BIAcore is used, it will be immediately apparent which combinations of monoclonal antibodies (Mabs) are suitable for a diagnostic assay.
4) Interactions are measured in real time, which allows kinetic rate constants and equilibrium affinity constants to be calculated (9-11). Knowledge of the affinity of viral antibodies is important for the proper design of immunoassays and for

understanding the molecular basis of viral epitope recognition and infectivity neutralization by antibodies (12, 13).

MAPPING OF ANTIGENIC SITES AND SELECTION OF ANTIBODY PROBES

For double antibody sandwich ELISA, it is necessary to select pairs of Mabs that are able to bind concurrently to the same antigen molecule because the epitopes they recognize are sufficiently distant on the surface of the antigen. Biosensor instruments are particularly well-suited for the rapid mapping of viral epitopes and for identifying which combinations of capturing and detector Mabs will give the best results in sandwich assays (14, 15).

In the present context, epitope mapping is the study of the relative position of epitopes on the surface of an antigen. This is usually done by competitive binding assays with pairs of Mabs. In such assays, two epitopes will be recognized as different only if they are far enough apart to allow simultaneous binding of the two Mabs. In many cases, however, Mabs directed against distinct but neighboring epitopes will be prevented from binding simultaneously to the antigen surface because of steric hindrance. The ability of pairs of Mabs to bind concurrently to the antigen can be measured by ELISA, but this requires labelling one of the two Mabs, for instance with biotin or with an enzyme. Competitive topographic mapping can be done much more conveniently with the BIAcore.

A two-site binding assay cycle with BIAcore consists of the following injection sequence: (1) undiluted hybridoma supernatant, containing the first Mab is injected over a rabbit antimouse IgG surface; (2) a nonspecific Mab at a high concentration to block any remaining free surface-binding sites; (3) the antigen; (4) the second Mab in undiluted hybridoma supernatant; and (5) regeneration of the sensor surface. After conditioning the surface with running buffer, it is ready for the next cycle. Typically, volumes of 5-10 µl are injected at a flow rate of 5 µl/min. The concentration of molecules binding to the sensor chip surface (expressed in resonance units or RU) is monitored continuously over time and is registered as a sensorgram. The Y-axis of the sensorgram is denoted as the RU signal whereas the time (in sec) is represented on the X-axis. Fig. 1 shows two sensorgrams of two-site binding assays performed after direct immobilization of a first capture Mab. The selectivity of the rabbit antimouse globulin surface eliminates the need for purification of the Mabs, and provides constant experimental conditions, since a new sample of the first Mab is used in each cycle.

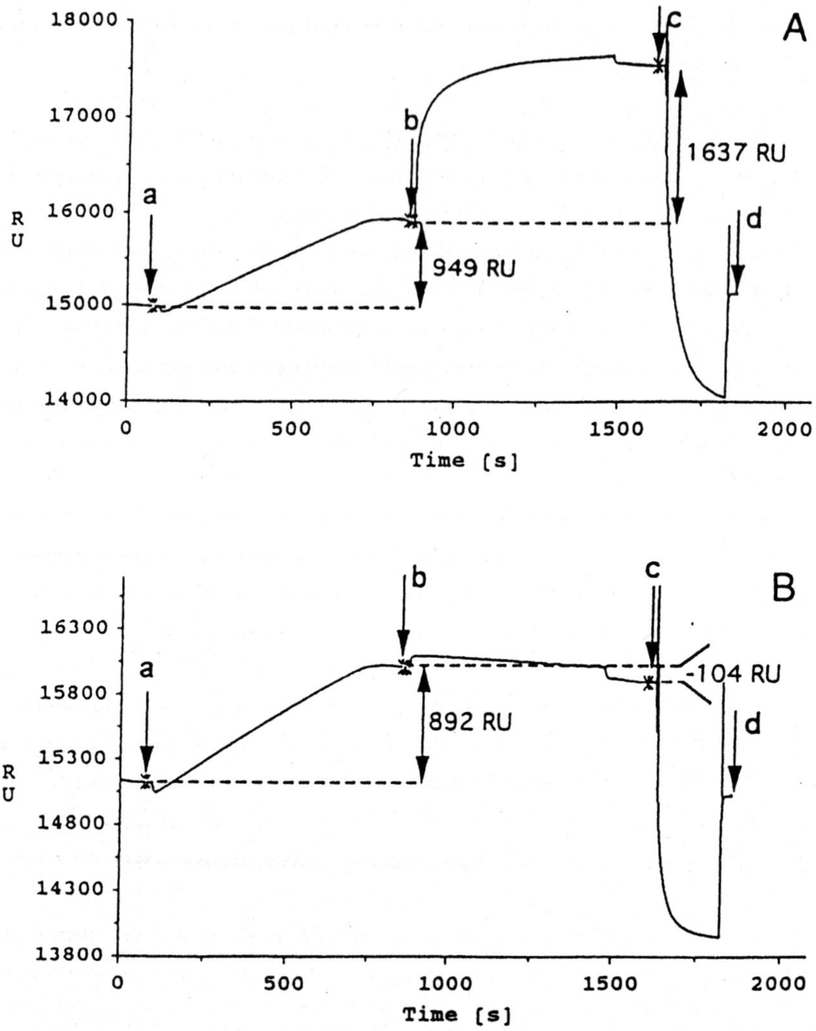

FIGURE 1

Sensorgrams corresponding to two-site binding assays. **(A)** The RU level at position **a** corresponds to the amount of capture Mab 1 (47P) covalently immobilized on the sensor chip. Phase **a** to **b** corresponds to the binding of 949 RU of TMVP. Phase **b** to **c** corresponds to the injection of Mab2 (151P). The positive response (1637 RU) indicates that Mab 151P does not overlap the epitope occupied by Mab 47P. Phase **c** and **d** corresponds to the injection of 10 µl of HCl 100 mM for the regeneration of the Mab1-immobilized surface by dissociation of noncovalently bound molecules. **(B)** The RU level at position **a** corresponds to the amount of captured Mab1 (47P) remaining linked to the matrix after the regeneration phase described above. Phase **a** to **b** corresponds to the binding of 892 RU of TMVP. Phase **b** to **c** corresponds to the injection of Mab2 (16P). The negative response (-104 RU) indicates that Mab 16P overlaps the epitope occupied by Mab 47P. Phase **c** to **d** corresponds to the regeneration phase. A flow rate of 5 µl/min was used during phases **a** to **d** (from Saunal and Van Regenmortel, 1995 (19), with permission).

Usually, antibodies can be captured and removed for about 100 assay cycles before the surface is consumed.

If a panel of 30 different Mabs is available, testing all combinations will involve 900 experiments. Changing the order of the formation of the complex can give rise to negative and positive binding for the same antibody pair. Therefore, a negative result should always be tested in the reversed order to verify the negative response (16). Proper evaluation of asymmetric interactions can be a source of valuable information (17). First, possible contributions of poor experimental design must be eliminated. For example, if one of a pair of noninterfering antibodies binds the antigen with low affinity or is added to the experiment at too low a concentration, the illusion of asymmetry can be created. Other causes are potentially more informative. Any type of heterogeneity among the antigen molecules can cause asymmetry if one of a noninterfering pair of antibodies is sensitive to that heterogeneity and the other is not. Heterogeneity may be caused by post-translational modifications, the formation of small aggregates, or the presence of conformational subpopulations.

During two-site binding assays, continuous dissociation and reassociation of one or more of the binding partners occurs. When the k_d of Mab1 is about $10^{-3}s^{-1}$, approximately half of the trapped antigen will become dissociated during the 10 min of Mab2 injection. On the dissociated antigen molecule, the part of the surface previously occupied by Mab1 may then bind to Mab2 and this will prevent reassociation of the antigen to Mab1. This results in a lower RU response. In view of the dynamic competition between different Mabs that can be observed with the BIAcore, this type of epitope mapping has been termed kinetic mapping. The reversible nature of antigen-antibody interactions can be visualized if the k_d of the immobilized first antibody lies in the region 10^{-2} to 10^{-4}. If the k_d value of Mab1 is larger than 10^{-2} s^{-1}, no meaningful data can be obtained since practically all antigen molecules will dissociate during the time frame of the experiment.

A necessary control in two-site binding assays with a monomeric protein consists in using the same antibody as both first and second antibody in order to check that no oligomers are trapped by the Mab1. A negative RU response indicates that the antigen actually dissociated from the capture Mab. After dissociating from the immobilized antibody, the antigen molecules are able to bind to the same free antibody introduced in the flow cell as Mab2 and this prevents antigen-Mab2 complexes from reassociating to the immobilized antibody.

A recent study of the epitopes of tobacco mosaic virus (TMV) protein demonstrates how mapping can be done in the case of viral antigens (18, 19). Antibodies elicited against TMV or dissociated TMV protein (TMVP) have been classified according to their ability to react with three different types of epitopes known as cryptotopes, neotopes and metatopes (20). Cryptotopes are epitopes that are buried inside the assembled virus particle and become accessible to antibodies only after dissociation of the virion. Neotopes are specific for the quaternary structure of the virion and are thus not present at the surface of monomeric subunits. Metatopes are present in both disssociated and polymerized forms of the viral protein. It has been possible to locate some cryptotopes in the TMVP molecule because the corresponding anticryptotope antibodies reacted with natural or synthetic fragments of the protein. This type of epitope mapping was not successful in the case of neotopes and metatopes because none of the corresponding antibodies reacted with any linear fragments of TMVP. Some information on the location of a limited number of neotopes and metatopes on TMV particles was previously obtained by immunoelectron microscopy. All examined antimetatope Mabs were found to react with the virion extremity containing the 5' end of the RNA whereas the antineotope Mabs reacted along the entire length of the virus particle (21).

Using the BIAcore, double antibody capture assays with viral subunits and binding stoichiometry calculations with virus particles made it possible to demonstrate the presence of neotope and metatope specificities on parts of the viral surface where they had not been found before by electron microscopy. When the stoichiometry of antibody binding to virus particles was approximately 10, it was concluded that the antibody interacted with one of the rod extremities. A stoichiometry of about 300 indicated that the antibody interacted with the entire length of the virus particle (18). Some metatopes were found to be located on the external surface A of the subunit (see Fig. 2) whereas some neotopes were identified at the extremity of the viral rod, i.e. on the polymerized surface E' of the subunit. The biosensor analysis thus made it possible to distinguish two types of antimetatope Mabs reacting with surfaces A and E of the subunit respectively, as well as two types of antineotope Mabs reacting with surfaces A' and E' of the virion (Fig. 2). The advantages of biosensor technology for analysing viral epitopes have been discussed elsewhere (13). In particular, the BIAcore makes it possible to demonstrate changes in the antigenic reactivity of viral subunits induced by the binding of a first Mab (21). When certain Mabs directed to TMV metatopes were immobilized on the sensor chip and allowed to capture dissociated viral subunits, a neotope conformation was induced in the protein as revealed by the subsequent binding of antineotope antibodies (22).

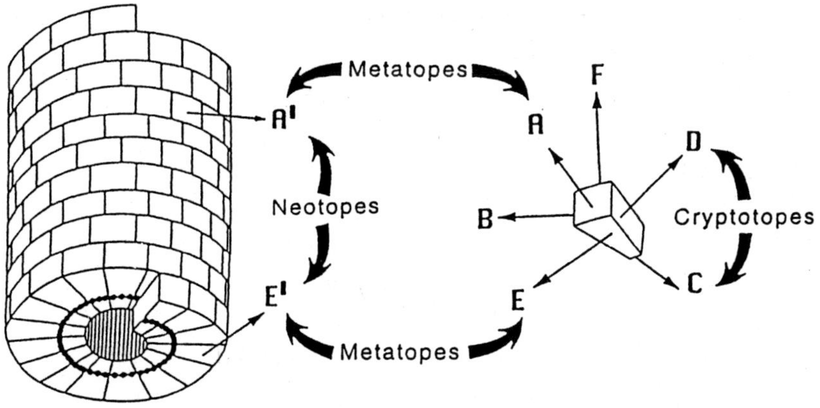

	4P	5V	6V	7V	16P	17V	18V	19V	25P	29V	42P	47P
A		+	+			+		+		+		
A'		+	+	+		+	+	+		+		
E					+				+			+
E'	+				+				+		+	+

	67P	79P	107P	151P	167P	175P	181P	188P	236P	249P	253P
A		+				+			+		
A'		+	+			+			+	+	+
E				+	+		+	+			
E'	+			+	+		+	+			

FIGURE 2

Schematic model of TMV protein subunits in monomeric form and in the virus particle, with an indication of which protein surface is recognized by each of 23 Mabs. The allocation was done on the basis of BIAcore mapping results (from Saunal and Van Regenmortel, 1995 (19), with permission).

A three dimensional drawing representing surfaces A and E of TMVP is presented in Fig. 3A. The results of two-site binding assays combined with previously obtained information regarding which surface, A or E, was recognized by each Mab (18), were used to establish the relative position of the epitopes on the map (Fig. 3B). The combining sites of the antibodies were assumed to cover a minimum surface of 600 Å^2 (23) in a circular footprint.

Pairwise interaction data give information only on the relative positions of epitopes since it is difficult to determine the orientation and the distances separating two nonoverlapping epitopes. As a result epitope maps are usually interpreted as functional "surface-like" maps (15, 17, 24, 25). In the TMV system, several factors made it possible to construct a schematic map depicting the actual

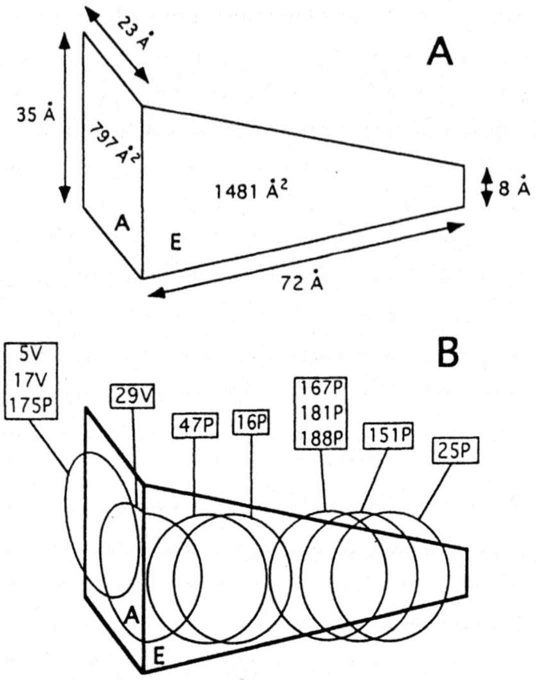

FIGURE 3

(A) Schematic model of surfaces A and E of TMVP. Dimensions were calculated from data described by Namba and Stubbs (48). (The virus particle is 300 nm long, 18 nm in diameter with a central hole of 0.4 nm and possesses 2130 identical protein subunits that form a helix of pitch 2.3 nm with 16 1/3 subunits in every turn.) (B) Schematic epitope map constructed from pairwise interaction data. Each antibody combining site was assumed to cover about 600 $Å^2$ in a circular footprint (from Saunal and Van Regenmortel, 1995 (19), with permission).

physical location of the epitopes on the antigen surface. Use was made of the known specificity for surface A or E of each Mab. The small size of the protein together with the observation that one Mab specific for surface A (29V) interfered with two Mabs specific for surface E (16P and 47P) allowed the unambiguous positioning of Mabs on the TMVP surface.

The epitope map shown in Fig. 3B was useful for understanding the mechanism of infectivity neutralization observed with some of the anti-TMV Mabs. Previous studies have shown that some Mabs reacting with surface E of the viral subunit were able to block the disassembly of TMV by ribosomes whereas others were devoid of such inhibitory capacity (26). No correlation could be established between the binding kinetics of the Mabs and their inhibitory capacity. However, the inhibitory Mabs (25P, 151P, 167P, 181P, 188P) were found to bind to the region of surface E closest to the central axis of the virus which is known to interact with the viral RNA. These data indicate that inhibitory Mabs act by sterically preventing the interaction between virus particles and ribosomes (19).

ANALYSIS OF VIRAL SYNTHETIC PEPTIDES

Several laboratories have used the biosensor technology in an attempt to optimize peptide probes used in viral diagnosis. The use of synthetic peptides corresponding to single epitopes increases the specificity of a diagnostic assay in the same way that Mabs do compared to polyclonal antiserum. In the case of human immunodeficiency virus (HIV), various linear and cyclic peptides corresponding to immunodominant epitopes of the gp 120 and gp 41 proteins have been studied using the peptides either as immobilized peptides, as free inhibitor peptides or as conjugated peptides. This type of study made it possible to quantitate antibody to specific epitopes in the serum of HIV-1 positive individuals enrolled in HIV-specific vaccine therapy trials (27).

In one particular study, monoclonal antibodies and rabbit polyclonal antibodies were elicited against a cyclic peptide representing a chimeric sequence of a consensus V3 loop of HIV-1 gp 120 (28, 29). When tested by ELISA and with the BIAcore, the antibodies cross-reacted extensively with the V3 regions of different HIV-1 strains and recognized the cyclic form of the homologous peptide better than its linear form. V3 peptides of different HIV-1 types were used to inhibit the binding of antibodies to the chimeric immobilized peptide (Fig. 4). In this way, peptide residues critical for antibody recognition could be identified. The chimeric peptide was also a good antigenic probe for detecting antibodies in the sera of patients infected with various HIV-1 strains (28). The ELISA titers of Mabs with respect to various HIV peptides correlated very well with the amount of competitor peptide needed to inhibit 50% of the binding of the Mab to the immobilized antigen present on the sensor chip. BIAcore binding data are therefore useful for selecting peptides to be used in diagnostic solid-phase immunoassays (30).

Recently, retro-inverso peptides which contain NH-CO peptide bonds instead of CO-NH bonds have been shown to be useful for mimicking viral epitopes (31-33). The binding properties of such peptidomimetics can be easily quantified with BIAcore. A retro-inverso peptide of the immunodominant loop 141-159 of the

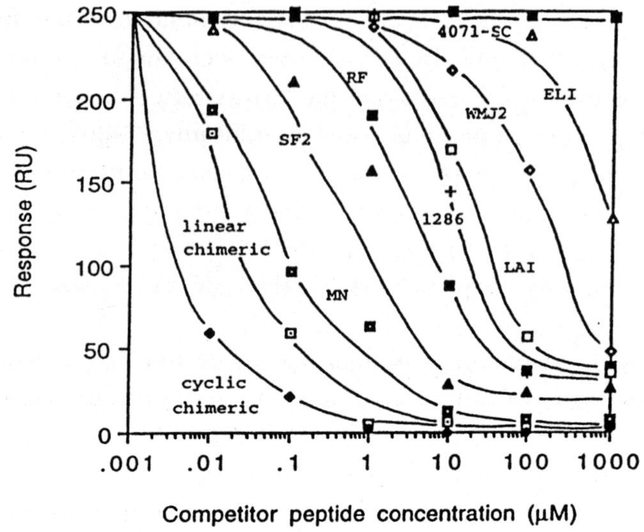

FIGURE 4

Capacity of increasing amounts of V3 peptides of different HIV-1 strains to inhibit the binding of a specific Mab to a chimeric V3 cyclic peptide immobilized on the sensor chip (from Richalet-Sécordel et al., 1994 (29), with permission).

VP1 protein of foot-and-mouth disease virus, when used as immunogen, produced higher serum antibody titers which appeared earlier after the start of immunization and lasted longer than those obtained with L-peptides. Analysis with the BIAcore showed that antibodies to retro-inverso peptides cross-reacted strongly with L-peptides and with virus particles, while antisera to VP1 protein and to virions also cross-reacted strongly with the retro-inverso peptides. In view of their increased resistance to proteolysis and high level of antigenic mimicry with viral epitopes, retro-inverso peptidomimetics should have considerable potential in viral immunology (33).

A commonly used method for localizing epitopes in a viral protein consists of identifying which peptide fragments of the protein are able to cross-react with antibodies elicited against the intact molecule. Any peptide of 10-15 residues that is able to bind to the antibodies is said to contain a continuous epitope. However, every residue in such a peptide is not a contact residue contributing to the binding energy of interaction. This can be demonstrated by comparing the binding of the wild type peptide with that of peptides presenting single residue modifications. Such comparisons are easily made with the BIAcore by measuring the rate constants characterizing the interaction of the antibody with the wild type and mutant peptides. When epitopes of TMV protein located in peptides 125-136 and 134-146 were analyzed in this way, it was found that only a few of the residues in these peptides contributed to the energy of interaction (34, 35). In most cases, the substitution in the mutated peptide affected the dissociation rate constant. This is illustrated in Fig. 5 which shows the on and off rates of a recombinant Fab fragment interacting with various immobilized peptides presenting single substitutions in the epitope region 134-146 of TMV protein (36). Very small changes in binding affinity can be measured with considerable precision, which is a prerequisite for analyzing the functional effect and thermodynamic implications of limited structural changes in interacting molecules.

The BIAcore is also useful for comparing the affinity of an antipeptide antibody for the free peptide and for the parent protein. This information is important since the usual purpose of raising antibodies to synthetic peptides is to obtain reagents that will cross-react strongly with the cognate protein and which could therefore be used as diagnostic reagents or potential synthetic vaccines (37). The similarity in kinetic binding constants observed when a cross-reactive antibody binds to the peptide and the protein is the best testimony that the peptide achieves a high degree of conformational mimicry with the cognate protein.

ANALYSIS OF VIRUS-ANTIBODY INTERACTIONS

The presence of large numbers of identical subunits at the surface of virus particles favors bivalent antibody binding. It is important to assess whether monovalent or bivalent binding occurs in any particular assay since the calculated affinity (better called avidity) may be 100-fold higher in the case of bivalent binding (38). This problem is most effectively overcome by using Fab fragments instead of intact antibody molecules in the binding studies.

The need to develop an optimal immobilization strategy for each virus under study can be obviated by capturing the virus by a first layer of immobilized antibody (39). Each virus particle will be captured by several antibody molecules and the binding is likely to be strong enough to withstand the dissociation and regeneration steps. The same virus surface could then be used repeatedly to analyze the binding properties of a series of Mabs used as second antibody.

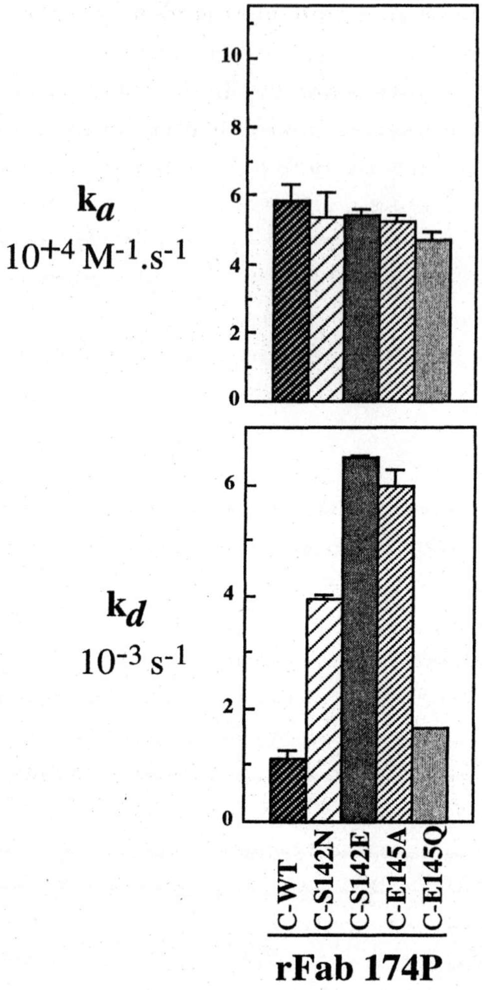

Sequence of peptide 134-151 of the coat protein of TMV

134 **142** **145** **151**	
C R G T G S Y N R **S** S F **E** S S S G L V	**C-WT**
N/E (position 142)	**C-S142N/E**
A/Q (position 145)	**C-E145A/Q**

FIGURE 5

On-rates (k_a) and off-rates (k_d) of a recombinant Fab (Mab 174P) reacting with immobilized wild type (WT) and mutant peptides corresponding to the region 134-151 of TMVP (from Chatellier et al., 1996 (36), with permission).

Knowledge of the affinity of viral antibodies is important for understanding the molecular basis of viral epitope recognition and of virus neutralization by antibodies (40). Methods used for determining the equilibrium affinity constant of viral antibodies have been reviewed recently (41). Virus-antibody binding data can be obtained by separating free antibody from virus-antibody complexes by ultracentrifugation and quantitating the free antibody by enzyme immunoassay (42). However, the binding constants of viral antibodies can be obtained much more easily by using the biosensor technology. On-rate (k_a) and off-rate (k_d) kinetic constants of the interaction between virus and antibody can be readily measured (10) and the equilibrium affinity constant K can be calculated from the ratio $k_a/k_d = K$.

The molecular basis of virus neutralization by antibodies is still poorly understood (43). Several authors have tried to establish a relationship between the affinity and neutralization capacity of Mabs specific for viral antigens. In some cases, it was found that antibodies neutralized viral infectivity in proportion to their affinity (44, 45) while in others no such relationship was found (46). Zolla-Pazner et al. (47) showed that there was a relationship between the affinity of human Mabs directed to the V3 loop of HIV-1 and their capacity to neutralize the virus. VanCott et al. (12) using the BIAcore showed that this relationship was due to corresponding changes in the dissociation rate constant rather than in the association rate constant of the antibodies. Since the kinetic profiles obtained with synthetic peptides and with purified recombinant gp 120 protein were similar, it was easier to screen many diverse isolates with peptides than with recombinant proteins.

REFERENCES

1. U. Jönsson and M. Malmqvist. Adv. Biosensors, 2, 291-336 (1992).

2. M. Malmqvist. Nature, 361, 186-187 (1993).

3. L.G. Fägerstam and R. Karlsson, In: Immunochemistry, C.J. Van Oss and M.H.V. Van Regenmortel eds., Marcel Dekker, Inc., New York, (1994) pp 949-970.

4. R. Granzow. Methods: A Companion to Methods in Enzymol., 6, 95-205 (1994).

5. M.H.V. Van Regenmortel. J. Immunol. Meth., 183, 3-182 (1995).

6. R. Cush, J.M. Cronin, W.J. Stewart, J. Maule, J. Molloy and N.J. Goddard. Biosensor Bioelectron, 8, 347-354 (1993).

7. J. Hodgson. Bio/Tech., 12, 31+35 (1994).

8. S.A. Darst, C.R. Robertson and J.A. Berzofsky. Biophys. J., 53, 533-539 (1988).

9. R. Karlsson, A. Michaelsson and L. Mattsson. J. Immunol. Meth., 145, 229-240 (1991).

10. J.L. Pellequer and M.H.V. Van Regenmortel. J. Immunol. Meth., 166, 133-143 (1993).

11. M. Malmqvist. J. Mol. Rec., 7, 1-7 (1994).

12. T.C. VanCott, F.R. Bethke, V.R. Polonis, M.K. Gorny, S. Zolla-Pazner, R.R. Redfield and D.L. Birx. J. Immunol., 153, 449-459 (1994).

13. M.H.V. Van Regenmortel, D. Altschuh, J.L. Pellequer, P. Richalet-Sécordel, H. Saunal, J.A. Wiley and Z.-L. G. Methods: A Companion to Methods in Enzymol., 6, 177-187 (1994).

14. E. Tarrab, L. Berthiaume, S. Grothe, M.M. O'Connor, J. Heppell and J. Lecomte. J. Gen. Virol, 76, 551-558 (1995).

15. G. Tosser, T. Delaunay, E. Kohli, J. Grosclaude, P. Pothier and J. Cohen. Virology, 204, 8-16 (1994).

16. L.G. Fägerstam, A. Frostell, R. Karlsson, M. Kullman, A. Larsson, M. Malmqvist and H. Butt. J. Molec. Recogn., 3, 208-214 (1990).

17. J.L. Daiss and E.R. Scalice. Methods: A Companion to Methods in Enzymol., 6, 143-156 (1994).

18. H. Saunal and M.H.V. Van Regenmortel. J. Immunol. Meth., 183, 33-41 (1995).

19. H. Saunal and M.H.V. Van Regenmortel. Virology, 213, 462-471 (1995).

20. M.H.V. Van Regenmortel, In: Immunochemistry of Viruses. II. The Basis for Serodiagnosis and Vaccines, M.H.V. Van Regenmortel and A.R. Neurath, eds., Elsevier, Amsterdam, (1990) pp 1-24.

21. I. Dore, E. Weiss, D. Altschuh and M.H.V. Van Regenmortel. Virology, 162, 279-289 (1988).

22. M.C. Dubs, D. Altschuh and M.H.V. Van Regenmortel. J. Chromatogr., 597, 127-148 (1992).

23. B.C. Braden and R.J. Poljak. FASEB J., 9, 9-16 (1995).

24. B. Johne, M. Gadnell and K. Hansen. J. Immunol. Meth, 160, 191-8 (1993).

25. E. Nice, J. Layton, L. Fabri, U. Hellman, A. Engstrom, B. Persson and A.W. Burgess. J. Chromatogr., 646, 159-68 (1993).

26. H. Saunal, J. Witz and M.H.V. Van Regenmortel. J. Gen. Virol., 74, 897-900 (1993).

27. T.C. VanCott, L.D. Loomis, R.R. Redfield and D.L. Birx. J. Immunol. Meth., 146, 163-176 (1992).

28. P.M. Richalet-Sécordel, A. Deslandres, S. Plaué, B. You, F. Barré-Sinoussi and M.H.V. Van Regenmortel. FEMS Immunol. Medical. Microbiol., 9, 77-88 (1994).

29. P.M. Richalet-Sécordel, G. Zeder-Lutz, S. Plaué, G. Sommermeyer-Leroux and M.H.V. Van Regenmortel. J. Immunol. Meth., 176, 221-234 (1994).

30. P.M. Richalet-Sécordel, F. Poisson and M.H.V. Van Regenmortel. Clin. Diagnost. Virol., 5, 111-119 (1996).

31. N. Benkirane, G. Guichard, M.H.V. Van Regenmortel, J.P. Briand and S. Muller. J. Biol. Chem., 270, 11921-11926 (1995).

32. G. Guichard, N. Benkirane, G. Zeder-Lutz, M.H.V. Van Regenmortel, J.P. Briand and S. Muller. Proc. Natl. Acad. Sci. USA, 91, 9765-9769 (1994).

33. S. Muller, G. Guichard, N. Benkirane, F. Brown, M.H.V. Van Regenmortel and J.P. Briand. Peptide Res., 8, 138-144 (1995).

34. D. Altschuh, M.C. Dubs, E. Weiss, G. Zeder-Lutz and M.H.V. Van Regenmortel. Biochemistry, 31, 6298-6304 (1992).

35. G. Zeder-Lutz, D. Altschuh, S. Denery-Papini, J.P. Briand, G. Tribbick and M.H.V. Van Regenmortel. J. Mol. Recogn., 6, 71-79 (1993).

36. J. Chatellier, N. Rauffer-Bruyère, M.H.V. Van Regenmortel, D. Altschuh and E. Weiss. J. Mol. Recogn., 9, 39-51 (1996).

37. M.H.V. Van Regenmortel, In: Immunochemistry, C.J. Van Oss and M.H.V. Van Regenmortel, eds., Marcel Dekker, New York, (1994) pp 277-301.

38. M.H.V. Van Regenmortel and G. Hardie. Immunochemistry, 13, 503-507 (1976).

39. M.C. Dubs, D. Altschuh and M.H.V. Van Regenmortel. Immunol.Letters, 31, 59-64 (1991).

40. E. Day. Advanced Immunochemistry. Wiley, New York (1990).

41. M.H.V. Van Regenmortel and A. Azimzadeh, In: Immunochemistry, C.J. Van Oss and M.H.V. Van Regenmortel, eds., Marcel Dekker, New York, (1994) pp 805-828.

42. A. Azimzadeh, J.L. Pellequer and M.H.V. Van Regenmortel. J.Mol.Recogn., 5, 9-18 (1992).

43. N.J. Dimmock. Neutralization of Animal Viruses. Springer-Verlag, Berlin (1993).

44. J.P.M. Langeveld, N.K.T. Back, P.J. Durda, J. Goudsmit and R.H. Meloen. J. Gen. Virol., 72, 2519-2526 (1991).

45. H.-P. Roost, M.F. Bachmann, A. Haag, U. Kalinke, V. Pliska, H. Hengartner and R.M. Zinkernagel. Proc. Natl. Acad. Sci., 92, 1257-1261 (1995).

46. L.E. Brown, J.M. Murray, D.O. White and D.C. Jackson. Arch. Virol., 114, 1-26 (1990).

47. S. Zolla-Pazner, J. Goudsmit and P. Nara. Semin. Virol., 3, 203-211 (1992).

48. K. Namba and G. Stubbs. Science, 231, 1401-1406 (1986).

IDENTIFICATION AND FUNCTION OF HOST DEFENSE CELLS BY FLOW CYTOMETRY

Carleton C. Stewart, Ph.D., Director
Laboratory of Flow Cytometry
Roswell Park Cancer Institute
Buffalo, New York

METHODOLOGY

Cytometry is the modern day term to describe instruments that are used to identify cells. The light microscope is a cytometer and when it is equipped with advanced optics, video camera and associated computers it is called an **image cytometer**. When cells are dispersed into a suspension which can flow in a moving stream for electronic interrogation, the instrument is called a **flow cytometer**. Whether a simple instrument that merely counts cells, such as an electronic (Coulter) particle counter or a more sophisticated laser-based instrument that also measures cellular fluorescence and can sort cells, often called a fluorescent activated cell sorter (FACS), a flow cytometer provides rapid high speed analysis of cells. Figure 1 shows a schematic representation of a flow cytometer. There are two kinds of flow cells. In both types, the cell stream merges with a sheath fluid at higher pressure which constricts the cell stream into a laminar flow. In the first type, known as stream in air, the cell stream and sheath fluid exit an orifice that is usually 70μ in diameter whereupon it intersects with the center of a laser beam. In the second type of flow cell the interrogation point is inside the flow chamber itself. The latter type has the advantage of providing better resolution because there is no refractive index change at the interrogation point. Because of its greater mass, droplet generation for sorting is more difficult.

As the cell enters the laser beam it scatters light in all directions. In the forward direction the light scatter is proportional to the cross sectional area of the cell and size information can be obtained. If the light scatter is measured at 90°, information on the internal structures, such as granules, can be measured. Cells can be stained with fluorescent dyes or with antibodies conjugated with fluorescent dyes. These dyes are excited by the laser and their emission is quantitatively measured by arrays of detectors whose optical path is 90° to the laser beam, shown in Figure 2. The quantitative values for each parameter are used to generate a computer file that can be evaluated by the user.

Once the properties of the cells of interest have been defined, that information is programmed into the flow cytometer. When a cell is detected, which meets the programmed

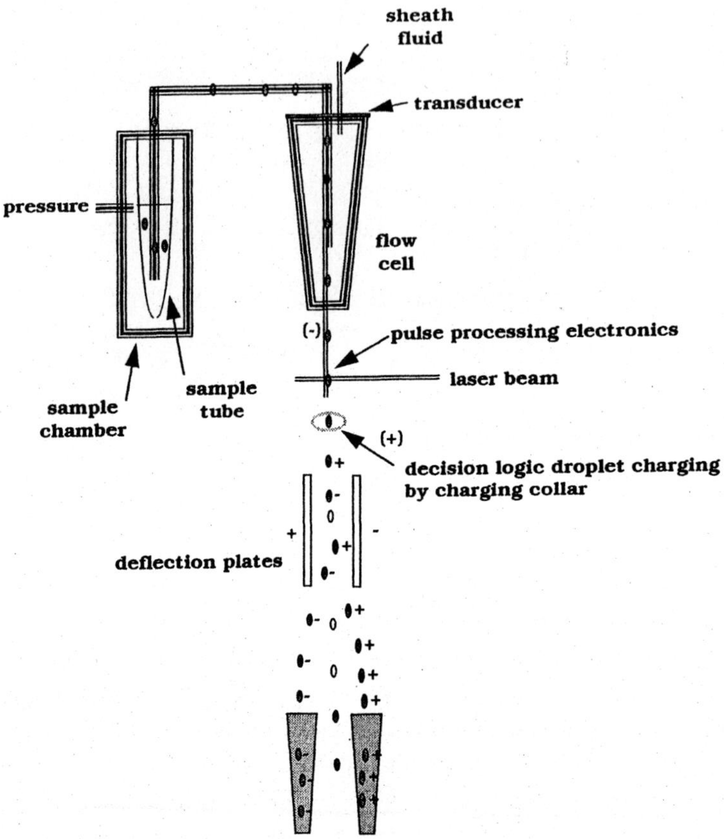

FIGURE 1

Schematic Representation of a FACS. Dispersed cells are suspended in suitable medium in a sample tube The sample tube is pressurized to force the cell suspension through sample tubing into the flow cell that contains a flowing stream of sheath fluid that is at a fixed pressure (10-50 psi). The sample pressure is increased so cells begin to flow into the sheath fluid that hydrodynamically focuses the cell stream at its intersection with the laser beam A piezoelectric transducer attached to the flow cell vibrates it at high frequency (25-100mH$_Z$) to create droplets that will contain individual cells for sorting. At the time of sorting, the charging collar is charged synchronously with the droplet containing the desired cell just as the droplet is breaking away from the cell stream. This effectively isolates the charge on the droplet as it falls between two deflection plates held at a high voltage (5000 or more volts). Positive droplets are attracted to the negative plate and negative droplets to the positive plate, effectively removing them from the uncharged droplet stream. The sorted droplets miss the plates and fall into collection tubes.

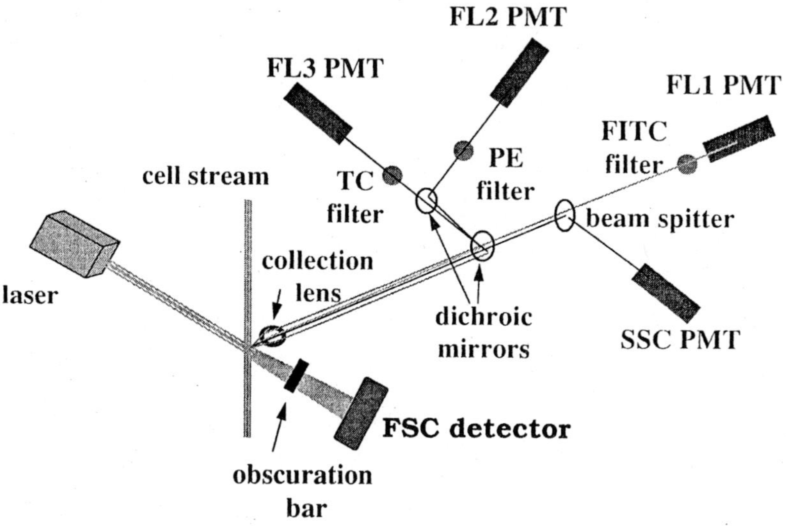

FIGURE 2

Schematic representation of optical light path. In this schematic representation, the optical layout for a five parameter flow is shown. Light scattered from the cell in the forward direction (7-22% off axis) is measured by a light sensitive diode detector. The side scatter and fluorescence measurements are made at 90% to the laser beam. A collection lens is used to collect light at the intersection of the cell stream and laser beam. The collected light is then passed through a series of dichroic mirrors and optical filters to photo multiplier (PMT) detectors. Dichroic mirrors are used to reflect appropriate wavelengths of light and transmit others. The scattered laser light and light from the fluorochrome with emission wavelengths less than 560 nM, pass through the first dichroic mirror while the longer wavelengths are reflected. A second dichroic is used to transmit the laser light into the PMT measuring side scatter and passes green light into the PMT measuring green fluorescence. A third dichroic mirror accepts the reflected fluorescence greater than 560nM and reflects orange light into one PMT and transmits red light into the other. Thus, the five parameters, forward scatter, side scatter, green, orange, and red fluorescence can be measured on each cell that passes the laser beam.

criteria, it is sorted. Sorting consists of vibrating the flow cells at very high frequency to create droplets.

The droplet containing the desired cell is electrically charged and falls through two high voltage deflection plates. If the droplet is charged positively it is deflected towards the negative plate. If it is charged negatively, it is deflected towards the positive plate. Thus, two populations can be sorted. Shapiro (1) or Given (2) have written an in-depth description of the cytometer operation.

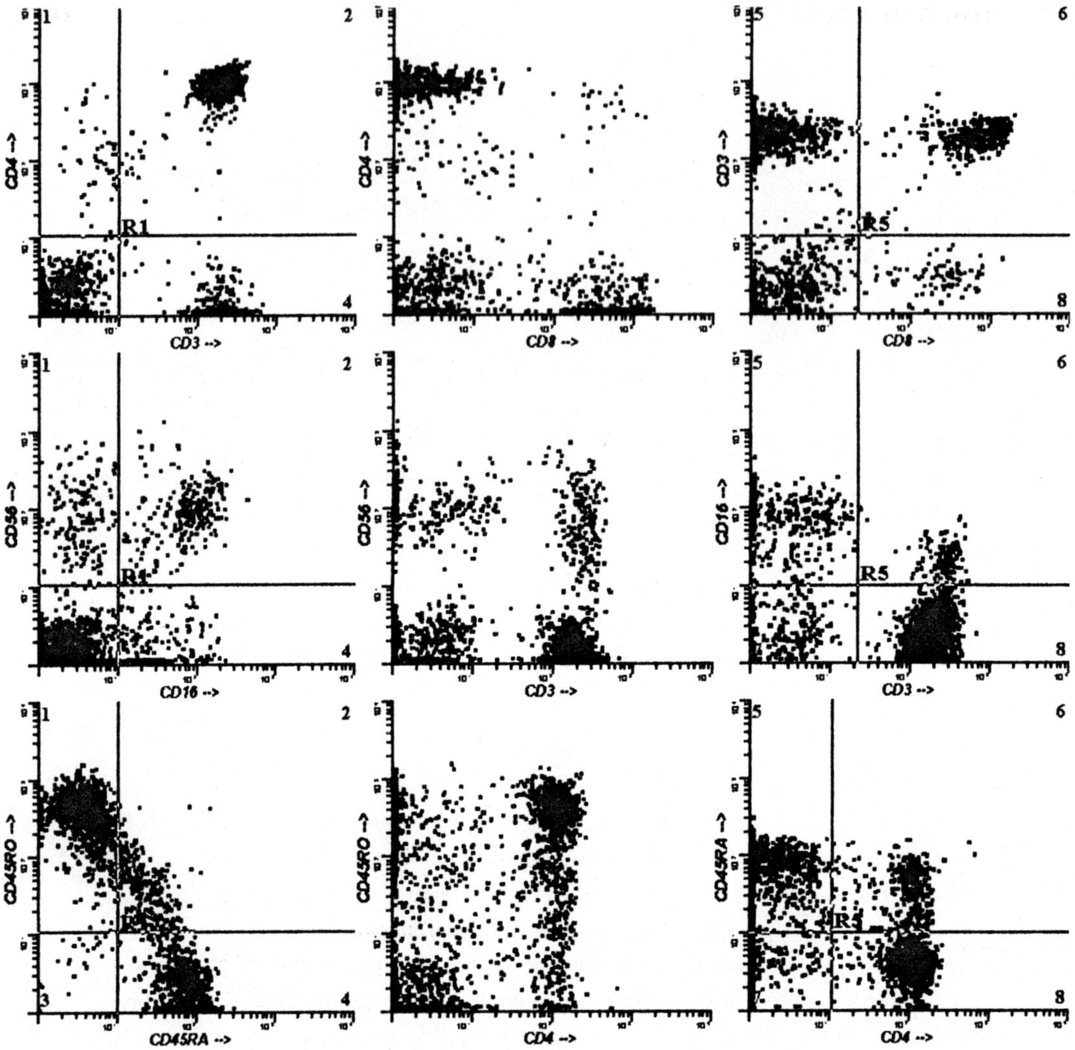

FIGURE 3

Lymphocyte Subsets: Human blood was stained with six separate combinations of antibodies to differentiation antigens expressed by lymphocytes. In the first row the combination of CD3 CD4 and CD8 produces a pattern that resolves the T-cell subsets CD3+CD4+CD8-, CD3+CD4+CD8+, CD3+CD4-CD8+ and CD3+CD4-CD8-. Note also the CD3-CD4+ monocytes in region 1 and the NK cell subset CD3-CD8+ in region 8. The second row shows the combination CD7, CD19 and CD2. T-cells are CD2+CD19-CD7+ shown in region 6. B-cells are found in region 1. NK cells are resolved as CD7+CD2- in region 5 and CD7-CD2+ in region 8. Those that are CD7+CD2+ cannot be resolved from T-cells. The third row resolves NK cells from T-cells using the antibody combination CD16, CD56 and CD3. T-cells are found in region 6 expressing CD16 and in region 8 cytotoxic T-cells expressing CD56 are clearly seen in the center view., NK cells can be divided into CD16-, shown in region 1 and CD16+ subset is shown in region 2. The fourth row shows that the γδ+ T-cells are CD4- (region 2), but some express CD8 (region 6). In the last two rows, the naive (CD45RA) and memory (CD45RO) subsets of CD+ and CD8 T-cells are shown in region 6.

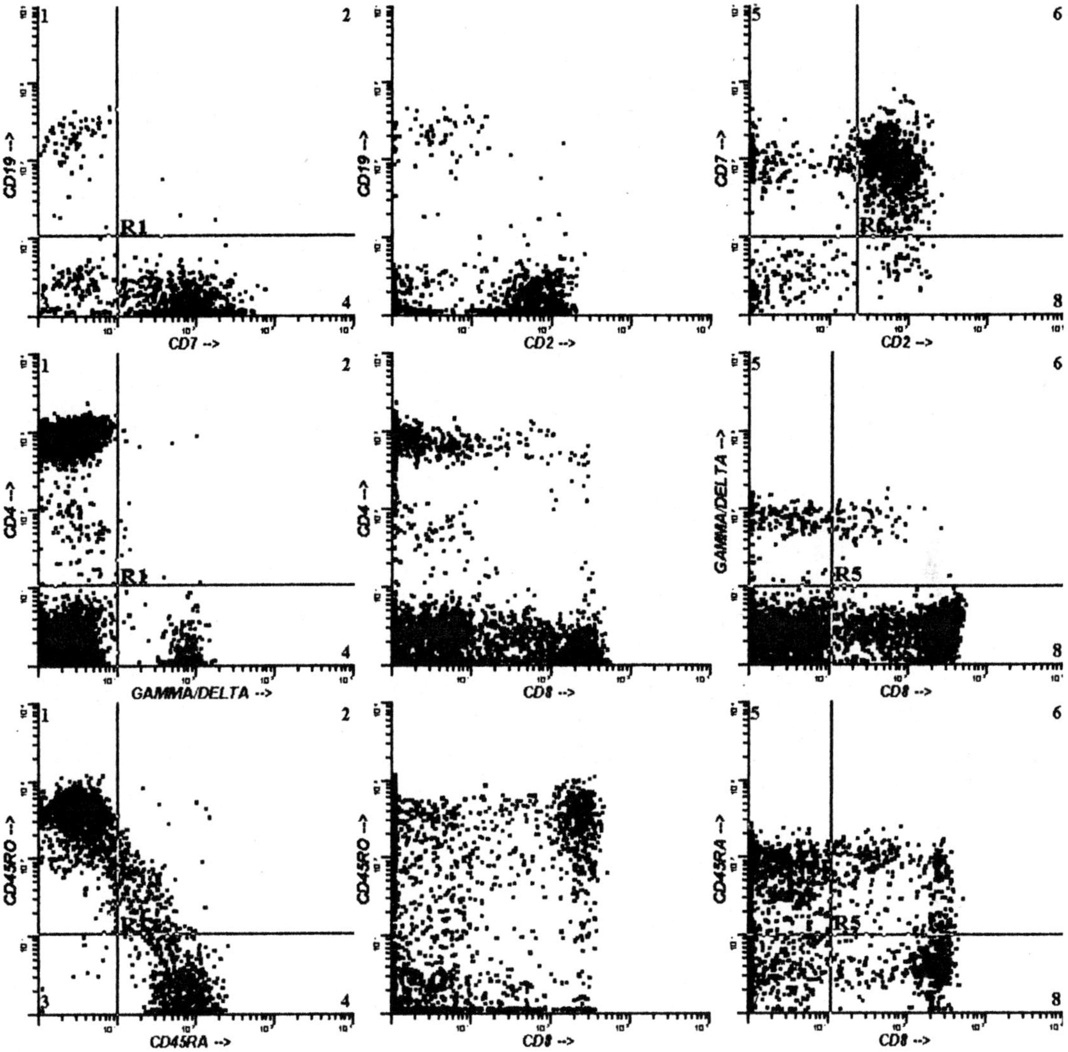

FIGURE 3 Continued

APPLICATIONS

The earliest measurements made by flow cytometers were the determination of the amount of DNA per cell. Cells were stained with dyes such as Ethidium Bromide or Propidium Iodide that quantitatively bind to the DNA. Today there are dyes that can be used to stain virtually every component found inside cells, such as RNA, protein, mitochondrial number and microtubular structures, and ion fluxes. Molecular probes (3) provides a complete supply of fluorescence reagents for performing these measurements. Darzynkiewicz, et al. have compiled a complete and comprehensive manual of procedures (4,5).

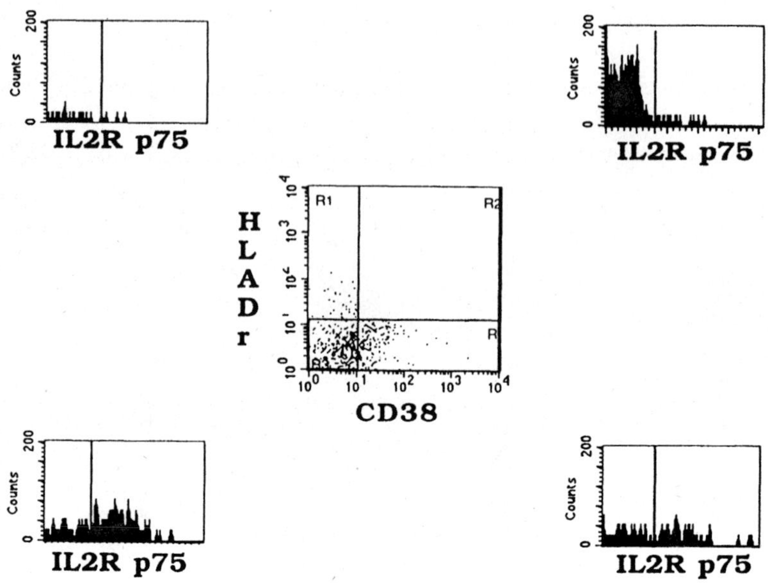

FIGURE 4

Cytokine Receptor Expression: Human bone marrow was stained with CD34, an antibody to hematopoietic progenitor cells, CD38, an anti-HLADr and an antibody to the p75 component of the 11.2 receptor. The bivariate plot in the center shows four populations of CD34+ cells: CD38-Dr+, CD38+Dr+, CD38+Dr and CD38-Dr-. Next to each quadrant is the univariate histogram illustrating the IL2R receptor expression for each population. Note that the highest expression is shown by the HLADr, CD38+/- populations.

The identification of lymphocyte subsets by immunophenotyping is a major application of flow cytometry. All cells contain structural proteins, proteins involved in cell communication, in cell metabolism and in cell locomotion or adherence. Antibodies made to these various proteins can be conjugated with different colored fluorochromes and used to quantify their expression. Some of these proteins do not change in amount on specific cell populations and are termed differentiation proteins. Other proteins are enhanced or suppressed depending on what the cell is doing; these are termed activation proteins. These cells can be stained with several antibodies to these proteins simultaneously for explicit identification and functional analysis.

To view flow cytometry data, one parameter is plotted against the other to produce a bivariate plot shown in Figure 3. Since each cell population has a unique repertoire of these proteins, they appear as distinct clusters in the bivariate plot creating a pattern for each combination of antibodies. Using combinations of three antibodies it is possible to resolve lymphocyte subsetsthat form clusters of cells resolved by the combinations. Not only is this a rapid enumeration technique, but these populations can be sorted for further evaluation.

NEW METHODS

A new methodology has been developed to identify intra-cellular proteins (antigens) (6). When combined with surface immunophenotyping to identify the cell, it is possible to determine what cells are producing what cytokines. If the cytokines are the words cells use to communicate with one another then their ears must be the cytokine receptors. Figure 4 shows a four antibody combination used to identify subsets of hematopoietic progenitor cells and evaluate their cytokine receptor repertoire.

CONCLUSIONS

In summary, flow cytometry can be used for the rapid identification of host defense cells and, using antibodies to functional proteins to determine the activation status of unique subsets. This offers the researcher a powerful approach to study the cellular interactions that occur as the host defends itself against disease.

ACKNOWLEDGEMENTS

The author acknowledges the contributions of Steven Rehbein, Sigrid Stewart and the Clinical Flow Cytometry Staff at RPCI. The research was supported by the New York State Department of Health and by grants NCI (5P30CA1605621) and (2RO1CA6020004) from the National Institutes of Health.

REFERENCES

(1) H.M. Shapiro, Practical Flow Cytometry, Wiley-Liss, Inc., New York, NY, 1995

(2) A.L. Givan, flow Cytometry: First Principles, Wiley-Liss, Inc., New York, NY 1992

(3) Molecular Probes, Inc., 4849 Pitchford Avenue, Eugene, OR 97402-9144

(4) C.C. Stewart, S.J. Stewart. In: Methods In Cell Biology, (Z. Darzynkiewicz, J. Robinson, H. Crissman, eds.) Academic Press, Inc., New York, 41:39-60, 1994.

(5) C.C. Stewart, S.J. Stewart. In: Methods In Cell Biology, (Z. Darzynkiewicz, J. Robinson, H. Crissman, eds.) Academic Press, Inc., New York, 41:61-79, 1994

(6) L.J. Picker, M.K. Singh, Z. Zdraveski, J.R. Treer, S.I. Waldrop, P.R. Bergstresser, V.C. Maino. Blood, 86:1408-1419, 1995

APPLICATION OF FLOW CYTOMETRY IN TRANSPLANTATION MEDICINE

T. Shanahan, Ph.D.
Department of Microbiology
State University of New York at Buffalo
Histocompatibility and Immunogenetics Division
Erie County Medical Center, Buffalo, New York 14215

ABSTRACT

Immunological rejection remains a major barrier to successful organ transplantation. Consequently, immunosuppressive intervention to prevent or control the rejection process renders the transplant recipient susceptible to infectious diseases. Flow cytometry has become a useful tool for monitoring immunological responses in transplant recipients. There are three areas of clinical transplantation immunology that may benefit from this technology. First, characterizing and classifying alloreactive antibodies by flow cytometry identifies high-risk donor and recipient combinations with greater precision. Second, the ability to detect subtle changes in the cellular components of the immune system cytometrically may facilitate the differential diagnosis of rejection, infection, and iatrogenic toxicity. Finally, the ease with which flow cytometry determines the adequacy or inadequacy of immunosuppressive therapy through T cell receptor analyses serves to maximize the beneficial effects of engraftment.

INTRODUCTION

Immunological responses to alloantigens involve a number of cellular and humoral events that may serve as useful monitors of the rejection process. Immune recognition begins with the interaction of cytotoxic and helper T cells with HLA class I and class II antigens, respectively. Subsequently, post-antigenic differentiation and activation occur with the expression of IL-2 receptors, transferrin receptors, and HLA-DR antigens. These events culminate with T cell expansion, infiltration, and mediation of graft destruction.

Additionally, T cell activation leads to the production of lymphokines that promote B cell differentiation and alloantibody formation. These antibodies are very important indicators of allogeneic immune responsiveness. Often their development is associated with irreversible graft rejection. Furthermore, their existence prior to engraftment as a result of pregnancies, transfusions, or a previous transplant, may predict the imminent destruction of engrafted tissue.

METHODS

ALLOANTIBODY ANALYSIS BY CYTOTOXICITY. The standard method used for the detection of alloreactive antibodies is the complement dependent cytotoxicity assay (1). In its most basic form, a patient's serum is combined with donor-derived lymphocytes that are representative of the allograft antigen profile. If donor-reactive antibodies are present, complement activation occurs with the addition of rabbit serum. This results in cell membrane destruction. Eosin serves as an indicator of cell membrane integrity where the uptake of dye suggests the presence of these antibodies. Exclusion of the dye on the other hand indicates their absence.

ALLOANTIBODY ANALYSIS BY FLOW CYTOMETRY. Flow cytometric analysis involves the detection of alloantibodies at the cell surface by indirect immunofluorescence (2). Sample preparation begins with the admixture of recipient serum and donor lymphocytes and follows with the addition of a fluorescein-conjugated secondary antibody such as goat anti-human immunoglobulin. T cell or B cell specific monoclonal antibodies, directly conjugated with red or orange fluorochromes, are used to subdivide and identify the lymphocyte populations. Sample analysis continues with the generation of a cytogram based on forward and right-angle light scatter properties of the cells. From this perspective, the lymphocyte population can be electronically selected for further analysis. Measurement of red or orange fluorescence enables the discrimination of lymphocyte subpopulations corresponding to the specificity of the monoclonal antibodies used. Analysis of green fluorescence serves as an indicator of donor-reactive recipient antibodies. With the use of donor peripheral blood leukocytes and appropriate gating techniques, monocyte antibodies are also detected. Furthermore, the use of the appropriate anti-human reagent determines the immunoglobulin class of the reactive antibody.

PHENOTYPIC ANALYSIS OF LYMPHOCYTES. Current approaches to lymphocyte phenotyping and subset analysis involve fluorochrome-conjugated monoclonal antibodies to various clusters of differentiation (CD) markers (3). By using different fluorochromes, fluorescein and phycoerythrin for example, the simultaneous analysis of multiple populations and dual marker analysis of individual populations become feasible. Lymphocyte markers commonly examined are the CD2 or CD3 pan T cell antigens, the CD4 marker of the T helper/inducer subset, and the CD8 marker of cytotoxic and suppressor T cells. The CD25 IL-2 receptor, the CD71 transferrin receptor, and T cell associated HLA-DR antigens serve as indicators of T cell activation.

Historically, measurements of lymphocyte populations entailed the subjective visual examination of fluorescent cells. However, statistical significance required the tedious analysis of a substantial number of cells. This was particularly challenging for dual marker analyses. With flow cytometry, however, objective analyses and vast enumerations of cells become feasible.

RESULTS

LIMITATIONS OF THE CYTOTOXICITY ASSAY. In the cytotoxicity assay, the presence of donor-reactive antibodies is predictive of graft rejection whereas their absence indicates a favorable graft outcome. However, these interpretations are based on two assumptions: first, that all antibodies detected in this assay will mediate graft rejection and second, that all antibodies capable of mediating graft destruction are detectable by this method. Neither assumption is without exception. For example, there are cases when alloantibody is not detected but engraftment results in sudden rejection. Conversely, there are cases when donor-reactive antibodies are detected but detrimental effects on the engrafted tissue are not observed. Possible explanations for these false negative results and false positive reactions are summarized in Table I. The false negative results encountered may be attributed to low titer antibodies that are below the threshold of detection in the cytotoxicity assay (4), non-cytotoxic antibodies (5) that are inefficient activators of the complement cascade, and antibodies to HLA class II determinants (6) that are not normally expressed by the T cell majority of the donor lymphocyte population. False positive reactions may be attributed to IgM class autoantibodies (7) that mimic IgG class alloantibodies, clinically irrelevant alloantibodies that recognize lymphocyte antigens not shared with the engrafted tissue (8), and immune complexes that are capable of initiating the complement cascade in a non-specific fashion.

TABLE I

Limitations of the Cytotoxicity Assay

False Negative Results Due To:	False Positive Reactions Due To:
• Low titer antibodies	• IgM class autoantibodies
• Non-cytotoxic antibodies	• Clinically irrelevant alloantibodies
• Class II antibodies	• Non-specific immune complexes

IMMUNOGLOBULIN CLASSIFICATION BY FLOW CYTOMETRY. The problem of false negative results in the standard cytotoxicity assay may be resolved to some extent by substituting B cells for T cells as the target population. B cells not only express HLA class II antigens that are ordinarily absent from T cells, but they also express class I antigens at a higher density (9). However, the use of B cells to enhance the sensitivity of the assay may sacrifice its specificity as B cells are also potent targets for autoreactive antibodies. Because these autoantibodies are predominately of the IgM class (10), one approach to this dilemma is the classification of the immunoglobulins by flow cytometry. In our studies, we examined patients who, by cytotoxicity methods, exhibited B cell reactivity without consistent T cell reactivity. When assayed by flow cytometry using the appropriate anti-immunoglobulin probes, we were able to discern those reactions attributable to IgG and IgM (Table II). Of 58 patients studied, 18 (31%) displayed IgM associated reactivity whereas 31 (53%) demonstrated reactivity mediated by immunoglobulins of the IgG class. Because reactions due to IgM are seldom regarded as clinically significant, we do not consider them a contraindication to transplantation in our program. Moreover, their presence has not been proven to be deleterious to graft survival. Determining the clinical significance of IgG reactions on the other hand is more challenging.

MULTITARGET ANALYSIS. One approach to determining the clinical significance of IgG class alloreactive antibodies is through multitarget analysis. By performing the appropriate electronic manipulations, monocyte reactivity can be measured in addition to lymphocyte reactivity. As a result, different patterns of reactivity are observed that correspond to different antigen specificities (Table III). For example, the ubiquitous distribution of class I HLA antigens suggests that antibodies reactive with T cells, B cells, and monocytes are class I

TABLE II

Distribution of IgG and IgM Donor-Reactive Antibodies in Transplant Patients

Immunoglobulin Class	Number (percentage)
IgM alone	18 (31)
IgG alone	31 (53)
mixed IgM and IgG	7 (12)
neither IgM nor IgG	2 (3)

TABLE III

Interpretation of Multitarget Flow Cytometric Analysis

T Cell	B Cell	Monocyte	Specificity
Positive	Positive	Positive	HLA Class I
Negative	Positive	Positive	HLA Class II
Negative	Negative	Positive	VEC
Positive	Negative	Negative	Non-HLA
Negative	Positive	Negative	Non-HLA

specific. Otherwise, B cell and monocyte reactivity, in the absence of T cell reactivity, implies the presence of class II specific antibodies. In either case, HLA antibodies are clinically significant and contribute to allograft rejection. Conversely, antibodies to T cells or B cells, without monocyte reactivity, are likely non-HLA in nature and hence clinically insignificant. Clinical significance, however, is not necessarily limited to the HLA system. For example, there exists a system of vascular endothelial cell (VEC) antigens that are expressed by monocytes as well (11). Although these antigens are not detected on lymphocytes, they are believed to be clinically relevant.

Previously, we reported three cases of renal transplantation performed in the presence of B cell antibodies and absence of T cell antibodies as determined by the cytotoxicity assay (2). Retrospective analysis by multitarget flow cytometry revealed three patterns of reactivity. One patient reacted with T cells, B cells and monocytes and another with B cells and monocytes without evidence of T cell recognition. Both patients, who exhibited HLA patterns of reactivity, experienced sudden, irreversible graft failures. The third patient was B cell reactive only. In this patient, the graft continues to function well. Although all three patients showed identical patterns of reactivity by the cytotoxicity method, flow cytometry was able to distinguish antibody specificities and predict the clinical significance of these reactions.

NONCYTOTOXIC ANTIBODIES. Despite the availability of various crossmatching methods, immediate graft rejections still occur in the absence of detectable cytotoxic antibodies. We encountered two such cases in our program. Although donor lymphocytes were not available for retrospective testing, we performed flow cytometric analyses using lymphocyte panels from unrelated donors. These studies indicated the presence of broadly reactive B cell antibodies in one patient and antibodies widely reactive with T cells and B cells in the other. Although direct correlation with donor reactivity was not possible, a significant level of risk could be attributed to these non-cytotoxic antibodies.

This experience, in addition to the fact that both cases involved renal regrafts, prompted us to examine the influence of previous graft history on alloantibody levels (Table IV). We examined patients with and without cytotoxic antibodies and subdivided each group based on previous transplant experience; those awaiting regrafts and those without previous transplants. In each case, we compared flow cytometric analysis with the cytotoxicity results. Essentially, those patients who exhibited cytotoxic antibodies, regardless of history, were also reactive by flow cytometry with 100% agreement. Similarly, of those patients awaiting a primary graft without cytotoxic antibodies, 95% were also non-reactive by flow cytometry. However, it was alarming to discover that of the patients who experienced a previous graft failure and showed no evidence of cytotoxic antibodies, 65% were in fact reactive by flow cytometry.

CLINICAL SIGNIFICANCE OF NON-CYTOTOXIC ANTIBODIES. Certainly, the importance of these non-cytotoxic antibodies is controversial. However, there is substantial direct and circumstantial evidence supporting their significance in regraft patients. In one study, acute regraft failures occurring in six renal patients were examined (12). When donor-specific cytotoxic antibodies were not detected, analyses were performed by flow cytometry. Using standard fluorescence channel shift criteria, donor directed T cell reactivity was evident in five of these patients. Furthermore, when a more sensitive fluorescence ratio was employed,

TABLE IV

Correlations Between Cytotoxicity and Flow Cytometry

Transplant Group	Cytotoxicity Results	n	Flow Cytometry Results: Negative	Flow Cytometry Results: Positive
Primary	Negative	19	18 (95%)	1 (5%)
	Positive	19	0	19 (100%)
Regraft	Negative	17	6 (35%)	11 (65%)
	Positive	14	0	14 (100%)

T cell reactivity was detected in all six cases. Parallel analyses of successful regraft cases in this study indicated the absence of donor-specific flow cytometric reactivity.

There are other data that support the clinical significance of these cytometrically defined antibodies. For example, one study showed that T cell antibodies detectable by flow cytometry, but undetectable by cytotoxicity, had no detrimental effect in patients receiving their first renal allograft (13). However, their presence was associated with significant graft loss in retransplant patients. When flow cytometric analyses are expanded to include T cells and B cells, these antibodies present an elevated risk of primary rejection as well (14). Yet, the risk is still more pronounced in regraft patients. Finally, other studies support the contention that B cell antibodies, as detected by flow cytometry, predict decreased graft survival rates in both primary and regraft patients (15).

In conclusion, antibodies detectable by flow cytometry but undetectable by cytotoxicity are clinically important. Furthermore, their presence is even more significant in regraft patients. However, these antibodies should not be considered absolute contraindications to transplantation but rather as substantial risk factors.

LYMPHOCYTE POPULATION ANALYSES. Two of the clinical challenges often faced in the post-transplant period is the differential diagnosis of acute rejection and viral infection and distinguishing steroid-resistant rejection from cyclosporin toxicity. There have been attempts to resolve these problems through the analysis of T cell activation markers. Because dual marker analyses of minute subpopulations are required, flow cytometry is useful in this regard.

Monitors of activation include IL-2 receptor-associated CD25 and transferrin receptor-associated CD71 coexpressed with the CD3 marker of T cells (Table V). These markers are not always successful in distinguishing rejection from infection (16,17). However, they do indicate an immunological assault and therefore differentiate episodes of rejection and infection from cyclosporin-induced nephrotoxicity in renal transplant recipients. Analysis of the CD71 marker appears to be particularly useful for monitoring anti-rejection therapy in heart transplant recipients (18). In these cases, failure of CD71 to return to base levels portends subsequent rejection episodes.

Studies of the expression of HLA-DR antigens on activated T cells have also been performed (19). Similar to other activation markers, the coexpression of DR with CD4 or CD8 provides little assistance with the differentiation of rejection and viral infection. More informative, however, are those studies that correlate increases in the CD4/CD8 ratio with active rejection and relative decreases in the CD4/CD8 index with infection (17,20). Additional studies have subdivided the CD4 helper/suppressor-inducer subset by CD45RA expression. In these cases, increases in the helper population, as determined by relative decreases in CD45RA expression, appear to be unique to the rejection process (17,21).

Certainly, this area of study remains controversial. However, as flow cytometry is applied to its full potential and more precise markers of cellular immune function are employed, more significant diagnostic patterns may emerge.

MONITORING IMMUNOSUPPRESSIVE INTERVENTION. Occasionally, there are rejection episodes that do not respond to standard forms of immunosuppressive therapy. These instances may require intervention with monoclonal antibodies directed against the T cell antigen receptor/CD3 complex. This reagent, commonly referred to as OKT3, is effective through at least two basic biological mechanisms. First, OKT3 antibody may react with the CD3 complex and, through opsonization or complement activation, promote clearance and depletion of the T cell population. Second, OKT3 antibody may crosslink CD3 complexes leading to the polarization and endocytosis of T cell antigen receptors thereby rendering the cells nonfunctional (22,23).

The adaptability of flow cytometry to the analysis of multiple determinants makes it a useful tool for monitoring the effects of OKT3 therapy. There are three important indicators of the effectiveness of OKT3 treatment: the CD3 T cell receptor marker, the CD2 pan T cell

TABLE V

Lymphocyte Marker Expression in Transplant Recipients

	Graft Rejection	Viral Infection	CyA Toxicity
CD3+CD25+	Increase	Increase	No Change
CD3+CD71+	Increase	Increase	No Change
CD3+DR+	Increase	Increase	No Change
CD4+DR+	No Change	No Change	No Change
CD8+DR+	Increase	Increase	No Change
CD4/CD8	Increase	Decrease	No Change
CD45RA-/45RA+	Increase	No Change	Not Reported

TABLE VI

Interpretation of Cell Markers in OKT3 Treatment

CD2	CD3	OKT3	Interpretation
-	-	-	Elimination
+	-	-	Modulation
+	+	-	Saturation
+	+	+	No effect

marker, and OKT3 epitope expression within the CD3 complex. This panel of markers provides a means of determining the effectiveness or ineffectiveness of treatment (Table VI). With the proper dosage, one should expect a 95% reduction in T cells 24 hours after initial OKT3 administration. A decline in all three markers reflects the elimination of T cells. It is not unusual for the CD2 marker to return during treatment (24). However, under favorable circumstances, the CD3 antigen should remain depressed. The emergence of the CD2 marker, without CD3 expression, indicates the presence of modulated T cells that are devoid of the antigen receptor complex. Coexpression of CD2 and CD3, on the other hand, suggests the presence of functional T cells and restoration of the rejection process. However, failure to detect the OKT3 epitope on the cell surface indicates that circulating monoclonal antibodies remain at a quantity sufficient for antigen receptor saturation (22). Finally, persistence of all three markers implies nonresponsiveness and development of immunological resistance to the murine monoclonal antibody (22,25).

SUMMARY

Flow cytometry offers many advantages for the study of humoral and cellular components of the immune system in the transplant patient. The ability to analyze alloreactive antibodies with greater sensitivity and specificity, helps avoid some forms of antibody-mediated rejection that are often unavoidable by standard methods. The ease by which flow cytometry characterizes major and minor lymphoid populations makes it a useful tool for monitoring cell-mediated events during the post-transplant period. The goals of laboratories utilizing flow cytometry for transplantation purposes are 1) to improve graft survival by defining high-risk donor and recipient combinations, 2) to promote graft accessibility by avoiding the unfair exclusion of low-risk combinations, and 3) to contribute to the diagnosis of rejection and management of anti-rejection therapy. Together, these applications serve to maximize the beneficial effects of transplantation while minimizing the adverse complications.

REFERENCES

1. K.K. Mittal, M.R. Mickey, D.P. Singal and P.I. Terasaki. Transplantation 6, 913-917 (1968).
2. T.J. Farley, T.C. Shanahan and W.R. Bartholomew. Transplantation 47, 535-537 (1989).
3. A.L. Jackson. Clinical Immunol Newsletter 15, 1-5 (1995).
4. H.M. Gebel and L.K. Lebeck. Clin Lab Med 11, 603-620 (1991).
5. S.S. Karuppan, S. Ohlman and E. Moller. Transplantation 54, 839-843 (1992).
6. T. Mohanakumar, C. Rhodes, G. Mendez-Picon, M. Goldman, C. Moncure and H. Lee. Transplantation 31, 93-95 (1981).
7. W.E. Braun. Hum Immunol 26, 245-260 (1989).
8. A. Ting, in Kidney Transplantation Principles and Practice, P.J. Morris, ed, W.B. Saunder, Philadelphia, (1988), pp. 183-213.
9. M.A. Pellegrino, M. Belvedere, A.G. Pellegrino and S. Ferrone. Transplantation 25, 93-95 (1978).
10. J.R. Chapman, C.J. Taylor, A. Ting and P.J. Morris. Transplantation 42, 608-613 (1986).
11. J. Cerilli, W. Bay and L. Brasile. Hum Immunol 7, 45-50 (1983).
12. M.C. Bonadonna, M.S. Orloff, O. Bronsther, M.C. Morris and M. Coppage. Hum Immunol 44, 165 (1995).

13. D.J. Cook, P.I Terasaki, Y. Iwaki, F. Strauss, et al, in Clinical Transplants 1988, P.I. Terasaki, ed, UCLA Tissue Typing Laboratory, Los Angeles, (1988), pp. 375-380.
14. R.J. Mahoney, K.A. Ault, S.R. Given, D.A. Leeber, et al. Transplantation 49, 527-535 (1990).
15. V.A. Lazda. Transplantation 57, 964-969 (1994).
16. M.S. Schubert, J.A. Motil, M.L. Cohen, R.M. Radvany, L.J. Koep and B.A. VanderWerf. Transplant Proc 18, 740-742 (1986).
17. C.F. Springgate, J.L. Hussey and D.H. Hayes. ASHI Quarterly 12, 21-23 (1988).
18. K. Hoshinaga, N.L. Wood, T.C. Wolfgang, T. Mohanakumar, et al. Transplant Proc 18, 743-744 (1986).
19. A. van Es, W.M. Baldwin, P.J. Oljans, H.J. Tanke, J.S. Ploem and L.A. van Es. Transplantation 37, 65-69 (1984).
20. M. Valeri, A. Piazza, N. Torlone, E. Poggi, D. Adorno and C.U. Casciani. Transplant Proc 23, 2275-2276 (1991).
21. P. Reinke, E. Fietze, W.D. Docke, F. Kern, R Ewert and H.D. Volk. Transplantation 58, 35-41 (1994).
22. R.B. Colvin and F.I. Preffer. Clin Lab Med 11, 693-714 (1991).
23. G. Goldstein. Transplant Proc 19, 1-6 (1987).
24. G.J. Zlabinger, D. Maurer, W. Ulrich, G. Meron, E. Pohanka and J. Kovarik. Transplant Proc 22, 1777-1778 (1990).
25. J.R. Thistlethwaite, J.K. Stuart, J.T. Mayes, F.P. Stuart, et al. Am J Kidney Dis 11, 112-119 (1988).

PART II

BACTERIAL AND PARASITIC INFECTIONS

CRITICAL ASSESSMENT OF GENE AMPLIFICATION APPROACHES ON THE DIAGNOSIS OF TUBERCULOSIS

Betty A. Forbes
Department of Pathology
State University of New York Health Science Center
Syracuse, New York 13210

ABSTRACT

The resurgence of tuberculosis prompted the development of a number of new options for the rapid laboratory diagnosis of *Mycobacterium tuberculosis* (MTB). One of the most promising and exciting methodologies has been the introduction of assays employing amplification technology to detect MTB directly in clinical specimens. Although amplification assays hold significant promise to improve the laboratory diagnosis of tuberculosis, the decision to perform or not perform these assays is complicated. The performance of in-house polymerase chain reaction (PCR) assays and two commercially-prepared assays, GenProbe's AMTD test and Roche's AMPLICOR PCR assay are reviewed. Regardless of the amplification format, all assays have decreased sensitivity with specimens that are acidfast bacilli (AFB) stain-negative. Data from these studies and others indicate possible potential pitfalls of amplification assays, those being sampling errors, the presence of substances in clinical specimens that inhibit the amplification assay, and clinical utility.

In light of these findings, the possible roles for these assays in the clinical microbiology laboratory are reviewed. In addition, factors such as cost, assay performance, etc. are discussed in order to facilitate the decision-making process concerning whether an amplification assay would be appropriate in a particular laboratory setting.

The recent resurgence in tuberculosis in the U.S. was a result of a number of complex factors such as HIV-infection, homelessness, and other socioeconomic factors (1). Further complicating the early outbreaks were frequent delays in laboratory reporting of smear, culture, and/or susceptibility results (2,3). As a result, a critical need for a more rapid diagnosis for tuberculosis became apparent.

In response to this demand, significant energy was expended to provide more rapid methods for the laboratory diagnosis of mycobacterial infections. For microbiologists, there is presently a number of options from which to select that can impact on mycobacteriology laboratory services, depending on their needs and resources. Probably one of the more promising and exciting options for rapid diagnosis are those employing amplification technologies. The last several years have witnessed a plethora of information regarding the ability of these technologies to diagnose mycobacterial infections. Most recently, an FDA-approved commercial assay used to detect *Mycobacterium tuberculosis* (MTB) directly in clinical specimens has become available and another, as of this writing, is close to receiving formal approval.

However, in parallel with these technological developments, the practice of laboratory medicine has become increasingly complex over the last several years. The decision process as to the extent and level of laboratory services for the diagnosis of mycobacterial infections, and in particular, the use of amplification assays, is no different in this regard. The clinical microbiologist is confronted with multiple issues such as accommodating increasing public health, regulatory, and therapeutic demands coupled with limited laboratory resources, i.e. of money and personnel. Importantly, the clinical utility of various laboratory procedures must also be assessed. Although amplification assays hold significant promise to improve the laboratory diagnosis of tuberculosis, the decision to perform or not perform these assays is complicated. To facilitate the decision making process as to the utility of these assays in a clinical microbiology laboratory setting, the performance and possible limitations of amplification assays, as well as other factors to consider will be discussed.

PERFORMANCE

Since PCR's discovery over 10 years ago, a number of other approaches to amplification has been introduced to detect MTB directly in clinical specimens. Some of these amplification formats include polymerase chain reaction (PCR), isothermic enzymatic amplification of MTB rRNA that is coupled to a hybridization protection assay, ligase chain reaction, and strand displacement amplification. A review of each of these formats that are presently at various stages of development is beyond the scope of this discussion. Some of the available published data on PCR assays, and on two commercial assays, GenProbe's Amplified Mycobacterium Tuberculosis Direct Test (GenProbe, San Diego, California) or AMTD test and Roche Diagnostic Systems', AMPLICOR™ (AMPLICOR; Roche Diagnostic Systems, Inc., Branchburg, New Jersey) test will be summarized. Based on my own experience with PCR and evaluations to date of other formats, I believe that key questions and issues surrounding these assays will be similar, regardless of the specific format as one attempts to develop their own strategies for the laboratory diagnosis of mycobacterial infections.

If attempting to draw definite conclusions following literature review of that fondly referred to as "home brew" PCR assays, one is immediately confronted with a major stumbling block in trying to compare findings from one study to another. Table I summarizes data from some of the studies published since 1989 using PCR to detect MTB directly in clinical specimens.

TABLE I

Studies evaluating in-house developed PCR assays for the direct detection of MTB in clinical specimens.[a]

AUTHOR	# OF SPECIMENS	SOLID MEDIA ONLY	OVERALL (%)		SMEAR		PRIMER REGION
			SENSITIVITY	SPECIFICITY	+ SENSITIVITY (%)	- SENSITIVITY (%)	
Brisson-Noël (4)	35	Y	100	100	100	100	65 Kda[b]
Pao (5)	284	Y	100	63	NA[c]	NA	65 Kda
Eisenach (6)	162	Y	98	99	100	NA	IS6110[d]
Cousins (7)	177	Y	NA	76	97	NA	MPB70[e]
Savić (8)	145	Y	95	93	66	NA	IS6110
Soini (9)	127	Y	70	100	NA	NA	32 Kda[f]
Veringa (10)	107	Y	94	75	NA	NA	IS6110
Anderson (11)	300	Y	95	98	NA	NA	IS6110
Clarridge (12)	>5,000	N	86	99	94	62	IS6110
Forbes (13)	734	N	87	98	88	78	IS6110 & PAB[g]
Kocagöz[h] (14)	78	Y	87	96	100	100	IS6110
Nolte (15)	313	Y	91	100	95	57	IS6110
Shawar (16)	384	N	74	95	90	53	IS6110
Yuen (17)	519	Y	96	100	100	92	PAB
Schluger (18)	65	NA	Active TB: 100 Any TB (past or present): 87	70 90	NA	NA	IS6110

[a] Reprinted with permission (44).
[b] Primers bracket a region of a gene that codes for a 65 kDa antigen of *M. tuberculosis*.
[c] Not available.
[d] Primers derived from the nucleotide sequence of the mycobacterial insertion sequence IS6110.
[e] Primers bracket a region of a gene that encodes for the MPB70 antigen of *M. tuberculosis*.
[f] Primers derived from the gene region encoding from the 32 kDa secreted mycobacterial protein.
[g] Primers bracket a region of a gene that encodes for a 38 kilodalton protein of *M. tuberculosis*.
[h] Clinical diagnosis with high suspicion was considered the gold standard.

While reviewing this literature, one is struck by the bewildering array of extraction procedures, primers, sample input for amplification, and cycling parameters. Even primers originating from the same area of the genome often target different regions. Thus, comparing the data of the various home brews is most difficult. Regardless, overall sensitivities and specificities range from 70% to 100%, and 63% to 100%, respectively. Note, however, that some evaluations published to date using the "home brew" PCR formats have compared their direct PCR results to culture on solid media; this laboratory practice is known to be significantly less sensitive than using a solid and liquid medium. Unfortunately, some do not even report their culture methods. Importantly, the sensitivity of in-house PCR assays for those clinical specimens that are acidfast bacilli (AFB) stain-negative is frequently significantly less than the sensitivity for specimens that are AFB stain-positive (table I).

As previously mentioned, a number of companies are in various stages of developing amplification assays for the direct detection of MTB in clinical specimens. Recently, GenProbe's AMTD test received FDA approval for use to detect MTB directly in AFB smear-positive respiratory specimens; Roche's AMPLICOR assay is currently under evaluation by the FDA. Both assays are commercially-available outside the United States. Some of the currently available data regarding the performance of these two assays are shown in table II.
In addition to these studies, several investigators have reported their findings in which these two commercial assays were compared head-to-head as well as with culture (table III).

LIMITATIONS

With few exceptions, it is clear that most "home brew" PCR assays as well as GenProbe's AMTD and AMPLICOR's demonstrate good sensitivity for specimens that are AFB smear-positive. However, for those specimens that are AFB stain-negative, the sensitivity, if data are presented, ranges between 50 to 100%, with most assays lying between 60 and 80%. Explanations for the lower sensitivity of amplification assays to detect organisms in smear-negative specimens point to possible pitfalls or limitations of these assays. Three major caveats for amplification assays must be considered: the presence of inhibitory substances, sampling error and the clinical usefulness. These caveats are in addition to obvious problems many laboratories have dealing with cross-contamination and sensitivity as so graphically reported by Noordhoek, et al. (32).

Many studies do not report the number of specimens that were inhibitory to the PCR assay. In those reports that have, rates of inhibition range from as low as 1% to as high as 21% (table IV).

TABLE II

Studies evaluating GenProbe's AMTD test and Roche's AMPLICOR assay for the direct detection of MTB in clinical specimens.

ASSAY	AUTHOR	# OF SPECIMENS	OVERALL SENSITIVITY/ SPECIFICITY	SMEAR + SENSITIVITY	SMEAR - SENSITIVITY
GenProbe	Jonas (19)	758	82/99	NR[a]	NR
	Abe (20)	135	92/100	100	77
	Miller (21)	594	91/99	100	73
	Pfyffer (22)	515	94/98	100	83
	Vlaspoler (23)	550	98.4/99.7	NR	NR
AMPLICOR	D'Amato (24)	985	66.7/99.6	94.7	55.3
	Wobeser (25)	1480	79/99	98	53
	Carpentier (26)	2073	86/98	94.5	74
	Moore (27)	1009	83/99	99	66
	Bergman (28)	956	79.4/99.6	97.6	42.9

[a]NR: Not reported.

TABLE III

Studies comparing the performance of GenProbe's AMTD test versus Roche's AMPLICOR assay for the direct detection of MTB in clinical specimens.

AUTHOR	# OF SPECIMENS	AMTD SENSITIVITY/ SPECIFICITY	AMPLICOR SENSITIVITY/ SPECIFICITY
Vuorinen (29)	256	82.6/100%	82.8/100%
Žolnir-Dovč (30)	281	94.3/99.1%	94.3/99.1%
Ichiyanma (31)	422	100/99.3%	97.3/98.9

TABLE IV

Inhibition rates of in-house PCR assays used for the direct detection of MTB in clinical specimens.

AUTHOR	SAMPLE SIZE	% INHIBITION
Anderson (11)	25 μL	Too problematic
	5 μL	3
Nolte (15)	5-20 μL	17
Kox (33)	10 μL	4
Clarridge (12)	25 μL	7.3-16.5
Soini (9)	UNK*	7.9
Forbes (13)	10 μL	16
Kirschner (34)	30 μL	21

*UNK: Unknown

Sampling error can also contribute to decreasing the sensitivity of amplification assays with smear-negative specimens. To illustrate, Jonas etal. (19) cultured routinely-processed respiratory specimens onto solid media and the remaining sediment was then tested by GenProbe's AMTD assay. Cultures that grew MTB were also quantitated. In her study, one third of cultures that grew MTB had fewer than 100 colony forming units (CFU)/ mL and only 54% of these were AMTD test positive. Twenty-four specimens that were AMTD-test negative and culture-positive for MTB were retested before and after the addition of MTB rRNA; 18 of the 24 specimens had ≤100 CFU/ML. Six specimens were positive and 12 were negative on retesting, whereas six were inhibitory to amplification. These data provide evidence that sampling errors and interfering substances can both contribute to decreasing amplification assays' sensitivity. Yajko and co-workers (38) also provided direct evidence for sampling error. In this study, serial dilutions of clinical specimens were performed and then cultured; the respective serial dilutions were also assayed by PCR. Results of this study showed that variable PCR results were obtained from sputum specimen dilutions of different patients when there were 24 to 42 CFUs present.

As a result of our high initial rates of inhibition with our direct assay, we circumvented the problem by detecting MTB directly in BACTEC 12B broth cultures (36). Clinical specimens were handled in a routine manner: all processed sediments were inoculated to Lowenstein-Jensen medium and a BACTEC 12B vial. An aliquot of 12B broth was removed when the GI on the BACTEC 460 instrument was ≥10 and stored at 4°C until assayed by PCR. We performed these assays twice a week. Prior to PCR, aliquots were boiled and then used for the amplification. For 490 BACTEC vials assayed by PCR in this manner, initial analysis of data revealed an overall sensitivity and specificity of 100% and 99.7%, respectively. Of significance, 42% of cultures positive for MTB were smear-negative yet all PCR results were positive. This PCR assay was introduced into routine clinical use over 3 years ago with well over 1000 additional BACTEC 12B vials with GIs ≥10 processed by PCR. To date, this assay has achieved >99.9% sensitivity and specificity when compared to culture. Importantly, we have decreased the time required for identification by probes in half in our laboratory. Laboratories that perform GenProbe assays directly off BACTEC bottles would also realize a savings in time, although may not be as great, since PCR can easily be performed as GIs ≥10 rather than 999. Greater savings in time could also be realized if PCR assays are run daily. This approach has also been applied using AMPLICOR's PCR assay with similar success.

Although the number of papers regarding the clinical utility of amplification based assays do not nearly approach those regarding their performance compared to smear and culture, there are some published data that indicate possible caveats regarding the clinical utility of these assays. Schluger and coworkers (18) assayed clinical specimens from 65 patients for the presence of MTB by PCR and then correlated results with AFB smear, culture, pathology, and clinical history. Based on their results, the authors concluded that PCR was unable to differentiate active disease from either prior, treated disease or asymptomatic infection. Walker (37) and Pfyffer (22) reported similar findings. Pfyffer et al. (22) tested more than 900 clinical specimens by Gen-Probes's AMTD test and compared results to smear and culture on solid and liquid media. Of particular interest were data summarizing the analysis of discrepant results between the GenProbe amplification test and culture. If discrepant results were obtained, clinical data was evaluated. Of note, a number of patients who had positive AMTD test results but negative cultures had a history of either MTB infection or exposure. As discussed by the authors, these GenProbe results could represent a false positive, latent infection due to MTB or an indication by developing active TB in the future.

An excellent study by Beige and others (38) reported similar findings when PCR results were prospectively compared with conventional laboratory diagnostic methods in addition to clinical findings. Although PCR appeared to exclude a diagnosis of tuberculosis based on their results, like Schluger, Beige detected MTB DNA in many samples from patients with a history of tuberculosis but no active disease.

Finally, Morris et al. (39) performed PCR on 86 BACTEC bottles with GIs ranging from 21 to 999 and then assessed the clinical utility of the PCR results for 10 patients infected with MTB by examining their medical records. It is noteworthy that nine of the 10 patients had received treatment before the BACTEC bottle turned positive. Four of these nine patients had

AFB stain-negative specimens. A positive PCR result on the original specimen would have been useful for two of the five smear-negative patients since these patients would have received treatment 10 to 13 days earlier. As pointed out by these authors, initial evaluations are a useful first step, but one shouldn't be misled into believing that a test is necessarily going to improve clinical management dramatically. Based on this study, amplification results may not change what is done for most smear-positive patients. Obviously, it would be most useful for smear-negative patients if a result led to earlier treatment. But even in this situation, many patients are appropriately being given treatment based on clinical family history, physical and x-ray findings and skin test results. These studies underscore the need for gathering of essential clinical data before final conclusions can be drawn as to the clinical utility of amplification assays.

POSSIBLE ROLES OF AMPLIFICATION ASSAYS

There are several possible roles for these amplification assays. First, all specimens submitted for mycobacterial stain and culture could be assayed by amplification.. However, few would argue that if this were to be the case, something would correspondingly have to be eliminated from the laboratory routine, particularly in light of the potential costs and time required to perform these assays,. Second, these assays could be performed on AFB stain-positive specimens only, particularly since these patients represent the greatest public health risk. However, a study by Yajko et al. (40) underscores the importance of evaluating each laboratory's set of circumstances regarding AFB stain-positive sputums. In this three-year retrospective study, the predictive value of an AFB smear for MTB was found to be quite high (92% for expectorated sputum specimens), despite the high prevalence of *Mycobacterium avium* complex in respiratory specimens. Thus, if the predictive value of an AFB smear for MTB is about 90% or greater in the geographic area served by your laboratory, this may not be an appropriate niche for amplification assays considering their expense compared to AFB stain.

Other possible roles for amplification assays might include monitoring the response to anti-tuberculous therapy and/or predicting relapses. Recently, Kennedy and co-workers analyzed serial sputum specimens obtained from 10 patients with smear-positive tuberculosis. Of interest, PCR detected two of four relapse cases earlier than culture. Another possible niche for amplification assays might be in specific clinical settings such as for the diagnosis of tuberculous meningitis or the diagnosis of tuberculosis in children where conventional methods are certainly less than optimal.

DEVELOPING A STRATEGY

From this overview, it becomes evident that there is no one straightforward answer as to delineating the niche for these assays in the clinical microbiology laboratory. Development of a strategy is complicated by a number of factors. First, the lack of a "gold standard" for the diagnosis of tuberculosis makes it particularly difficult to evaluate any new methodology. But, the issues of sensitivity, specificity, turn-around time and expertise required must all be factored into the decision-making process. Also, the issue of cost must be reckoned with. Terms such as health care reform, merger, downsizing, cost constraints have become all too familiar to those of use involved in health care. Importantly, given the expense of these assays, the decision by on the assay's performance, but on the assay's impact on patient management. This final issue surrounding the cost of amplification assays, the cost, have just begun to be examined. Conly and coworkers (42) at Toronto Hospital performed a cost effective analysis of three diagnostic strategies for MTB. Following routine decontamination and concentration procedures, conventional mycobacteriology methods for culture and identification, i.e. smear, BACTEC culture, and NAP test for identification of MTB, were compared to results of either 1) a BACTEC culture with nucleic acid hybridization for identification, or 2) PCR using Roche's AMPLICOR assay on AFB smear-positive specimens, or 3) PCR using Roche's AMPLICOR assay on AFB smear-negative and -positive specimens to determine the costs per isolation day prevented. A number of parameters were included in their model such as isolation costs, the number of days to the first positive culture, number of admissions per month with a positive AFB smear, and number of days in isolation. Based on their model, they determined that the use of AMPLICOR on smear-positive specimens was the most cost-effective strategy. However, the authors cautioned that varying volumes and the need for isolation can dramatically alter the cost per isolation day. Another study by Fraser and Wilkins (43) examined the impact of amplification on the diagnosis of tuberculosis in six smear-negative patients diagnosed with tuberculosis. Of these 6 patients, 2 were culture and PCR positive for MTB, one was culture-positive but PCR-negative for MTB, while 3 patients were culture- and PCR-negative for MTB. Two of these 6 patients suffered a significant delay in diagnosis with the requirement for invasive and costly investigations yet only one was positive by PCR. Based on these findings, no cost benefit was demonstrated Keep in mind, however, that the number of smear-negative specimens in this study was small and that 50% of these patients were treated empirically following the clinical evaluation.

As a result, a clinical microbiologist is faced with enormous pressures to provide more rapid diagnosis of mycobacterial infections yet exercise cost restraints. It behooves all of us in these times to seriously consider both the financial and clinical impact of these assays before implementing any of these assays. Importantly, each laboratory faces a different set of circumstances and must therefore wrestle with issues such as appropriate utilization and cost-effectiveness when considering use of the new assays to diagnose *Mycobacterium tuberculosis* infections. If one is to be involved in the laboratory diagnosis of mycobacterial infections, one must take into consideration the numerous factors including the sensitivity, specificity, clinical utility, prevalence, cost and resource prior to choosing a strategy, particularly as the new molecular diagnostic methods are introduced into the marketplace. In light of the resurgence of tuberculosis, there is an immediate mandate for all of us to re-examine our present practices in the mycobacterial laboratory with a focus on improving our methods' sensitivity and decreasing our turn-around time as much as your resources in time, monies, and personnel will allow.

The resurgence of tuberculosis in the United States and its subsequent control is a major challenge for all of us involved in health care. This challenge has resulted in an explosion of promising, creative and new methodologies to afford better diagnostic laboratory methods. These efforts encompass a variety of alternatives; the list of which will continue to grow and demand continued re-evaluation of our laboratory practices in the mycobacteriology laboratory. Nevertheless, these are exciting times because of the promise that many of these new methods hold for the rapid diagnosis of mycobacterial infections.

REFERENCES

1. K. Brudney and J. Dobkin. Am. Rev. Respir. Dis. 144, 745-749 (1992).

2. R.E. Huebner, R.C. Good, and J.I. Tokars. J. Clin. Microbiol. 31: 771-775 (1993).

3. F.C. Tenover, J.T. Crawford, R.E. Huebner, L.J. Geiter, C.R. Horsburgh, Jr., and R.C. Good. J. Clin. Microbiol. 31: 767-770 (1993).

4. A. Brisson-Noël, C. Aznar, C. Chureau, S. Ngyen, C. Pierre, M. Bartoli, R. Bonete, G. Pialoux, B. Gicquel, and G. Garrigue. Lancet ii: 1069-1071 (1989).

5. C.C. Pao, B. Yen, J.-B. You, J.-S. Maa, E.H. Fiss, and C.H. Chang. J. Clin. Microbiol. 28: 1877-1880 (1990).

6. K.D. Eisenach, M.D. Sifford, M.D. Cave, J.H. Bates, and J.T. Crawford. Am. Rev. Respir. Dis. 144: 1160-1163 (1991).

7. D.V. Cousins, S.D. Wilton, B.R. Francis, and B.L. Cow. J. Clin. Microbiol. 30: 255-258 (1992).

8. B. Savić, U. Sjöbring, S. Alugupalli, L. Larsson, and H. Miörner. J. Infect. Dis. 166: 1177-1180 (1992).

9. H. Soini, M. Skurnik, K. Liippo, E. Tala, and M.K. Viljanen. J. Clin. Microbiol. 30: 2025-2028 (1992).

10. E. Veringa, B. van Harsselaar, and P. Hermans. J. Microbiol. Methods 16: 139-147 (1992).

11. Å.B. Anderson, S. Thybo, P. Godfrey-Fausett, and N.G. Stoker. Eur. J. Microbiol. Infect. Dis. 12: 922-927 (1993).

12. J.E. Clarridge III, R.M. Shawar, T.M. Shinnick, and R.B. Plikaytis. J. Clin. Microbiol. 31: 2049-2056 (1993).

13. B.A. Forbes and K.E.S. Hicks. J. Clin. Microbiol. 31: 1688-1694 (1993).

14. T.Kocagöz, E. Yilmaz, S. Özkara, S. Kocagöz, M. Hayranm M. Sachedeva, and H.F. Chambers. J. Clin. Microbiol. 31: 1435-1438 (1993).

15. F.S. Nolte, B. Metchock, J.E. McGowan, Jr., A. Edwards, O. Okwumabua, C. Thurmond, P.S. Mitchell, B. Plikaytis, and T. Shinnick. J. Clin. Microbiol. 31: 1777-1782 (1993).

16. R.M. Shawar, F.A.K. El-Zaatari, A. Nataraj, and J.E. Clarridge. J. Clin. Microbiol. 31: 61-65 (1993).

17. K.Y. Yuen, K.S. Chan, C.M. Chan, B.S.W. Ho, L.K. Dai, P.Y. Chau, and M.H. Ng. J. Clin. Pathol. 46: 318-322 (1993).

18. N.W. Schluger, D. Kinney, T.J. Harkin, and W.N. Rom. Chest 105: 1116-1121 (1994).

19. V. Jonas, M.J. Alden, J.I. Curry, K. Kamisango, C.A. Knott, R. Lankford, J.M. Wolfe, and D.F. Moore. J. Clin. Microbiol. 31: 2410-2416 (1993).

20. C. Abe, K. Hirano, M. Wada, Y. Kazumi, M. Takahashi, Y. Fukasawa, T. Yoshimura, C. Miyagi, and S. Goto. J. Clin. Microbiol. 31: 3270-3274 (1993).

21. N. Miller, S.G. Hernandez, and T.J. Cleary. J. Clin. Microbiol. 32: 393-397 (1994).

22. G.E. Pfyffer, P. Kissling, R. Wirth, and R. Weber. J. Clin. Microbiol. 32: 918-923 (1994).

23. F. Vlaspoler, P. Singer, and C. Roggeveen. J. Clin. Microbiol. 33: 2699-2703 (1995).

24. R.F. D'Amato, A.A. Wallman, L.H. Hochstein, P.M. Colaninno, M. Scardamaglia, E. Ardila, M. Ghouri, K. Kim, R.C. Patel, and A. Miller. J. Clin. Microbiol. 33: 1832-1834 (1995).

25. W.L. Wobeser, M. Krajden, J. Conly, H. Simpson, B. Yim, M. D'Costa, M. Fuksa, C. Hian-Cheong, M. Patterson, A. Phillips, R. Bannatyne, A. Haddad, J.L. Brunton, and S. Krajden. J. Clin. Microbiol. 34: 134-139 (1996).

26. E. Carpentier, B. Drouillard, M. Dailloux, D. Moinard, E. Vallee, B. Dutilh, J. Maugein, E. Bergogne-Berezin, and B. Carbonnelle. J. Clin. Microbiol. 33: 3106-3110 (1995).

27. D.F. Moore and J.I. Curry. J. Clin. Microbiol. 33: 2686-2691 (1995).

28. J.S. Bergman and G.L. Woods. J. Clin. Microbiol. 34: 1083-1085 (1996).

29. P. Vuorinen, A. Miettinen, R. Vuento. and O. Hällström. J. Clin. Microbiol. 33: 1856-1859 (1995).

30. M. Žolnir-Dovč,, M. Poljak, K. Seme, A. Rus, and T. Avšič -Županc. Infection 23: 216-220 (1995).

31. S. Ichiyama, Y. Iinuma, Y. Tawada, S. Yamori, Y. Hasegawa, K. Shimokata, and N. Nakashima. J. Clin. Microbiol. 34: 130-133 (1996).

32. G.T. Noordhoek, A.H.J. Kolk, G. Bjune, D. Catty, J.W. Dale, P.E.M. Fine, P. Godfrey-Fausett, S.N. Cho, T. Shinnick, S. B. Svenson, S. Wilson, and J.D.A. van Embden. J. Clin. Microbiol. 32: 277-284 (1994).

33. L.F.F. Kox, D. Rhienthong, A.M. Miranda, N, Udomsantisuk, K. Ellis, J. van Leewen, S. van Heusden, S. Kuijper, and A.H.J. Kolk. J. Clin. Microbiol. 32: 672-678.

34. P. Kirschner, J. Rosenau, B. Springer, K Teschner, K. Feldmann, and E.C. Bottger. J. Clin. Microbiol. 34: 304-312 (1996).

35. D.M. Yajko, C. Wagner, V.J. Tevere, T. Kocagöz, W.K. Hadley, and H.F. Chambers. J. Clin. Microbiol. 33: 1944-1947 (1995).

36. Forbes, B.A. and K.E. Hicks. J. Clin. Microbiol. 32: 1725-1728 (1994).

37. D.A. Walker, I.K. Taylor, D.M. Mitchell, and R.J. Shaw. Thorax 47: 690-694 (1992).

38. J. Beige, J. Lokies, T. Schaberg, U. Finckh, M. Fischer, H. Mauch, H. Lode, B. Köhler, and A. Rolfs. 33: 90-95 (1995).

39. Morris, A., L. Reller, and B. Devlin. J. Clin. Pathol. 47:190 (1994).

40. D.M. Yajko, P.S. Nassos, C.A. Sanders, J.J. Madej, and W.K. Hadley. Clin. Infect. Dis. 19: 334-336 (1994).

41. N. Kennedy, S.H. Gillespie, A.O. Saruni, G. Kisyombe, R. McNerney, F.I. Ngowi, and S. Wilson. J. Infect. Dis. 170:713-716 (1994).

42. J. Conly, S. Krajden, M. Krajden, R. Bannatyne, W. Wobeser, and the Toronto Rapid Diagnostics Consortium. Abstract D93, p.83. *In* Program and Abstracts of the 35th Interscience Conference on Antimicrobial Agents and Chemotherapy, American Society for Microbiology, Washington, D.C. (1995).

43. I.S. Fraser and E.G.L. Wilkins. Abstract D94, p.83. *In* Program and Abstracts of the 35th Interscience Conference on Antimicrobial Agents and Chemotherapy, American Society for Microbiology, Washington, D.C. (1995).

44. Forbes, B.A. Clin. Microbiol. Newsletter 17:145-150. 1995.

BORRELIA BURGDORFERI INFECTION: CLINICAL DIAGNOSTIC TECHNIQUES

P.K. Coyle
Department of Neurology
School of Medicine
State University of New York at Stony Brook
Stony Brook, New York 11794

ABSTRACT

Borrelia burgdorferi is a tick-borne spirochete and the etiologic agent of Lyme disease. This pathogen now accounts for 91% of vector-borne infections in the United States, and from a public health viewpoint is one of our major emerging infectious disorders. Specific properties of *B.burgdorferi* have resulted in diagnostic problems, including the lack of a readily available laboratory assay to detect active infection. Most laboratory testing for Lyme disease relies on serologic documentation of prior exposure to the agent. However, such testing detects asymptomatic infections, and does not detect seronegative infections. In addition, antibody tests for Lyme disease are not standardized. Cases of Lyme disease are both underdiagnosed and overdiagnosed. This review will discuss the spirochetal properties which contribute to diagnostic difficulties, will discuss current laboratory diagnostic tests, including serology and detection of *B.burgdorferi* DNA, and will discuss diagnostic tests in development, including recombinant-based serologic assays and detection of *B.burgdorferi* antigens.

INTRODUCTION

Lyme disease is a systemic bacterial infection due to a tick-borne spirochete, *Borrelia burgdorferi* (1-3). Almost all human infections are transmitted by tick bite, during sustained blood feeding by an infected genus Ixodes hard-body tick. It has been estimated that up to 50% of infections are asymptomatic and never lead to disease (4,5). Clinical syndromes have been associated with early local infection, early disseminated infection, and late disseminated infection (Table I). Although this is a systemic condition, Lyme disease targets to particular body organs to cause characteristic skin, joint, heart, eye, and nervous system involvement. Syndromes which occur in early infection will spontaneously remit even without antibiotic treatment. Syndromes which occur in late infection will slowly worsen unless antibiotic treatment is given.

TABLE I

Clinical Stages of Lyme Disease

Early local infection	Early disseminated infection	Late disseminated infection
• Erythema migrans (EM) • Mild viral-like illness	• Multifocal EM • Lymphocytoma (Europe) • Cardiac disease • Arthritis • Hepatitis • Conjunctivitis • Neurologic disease	• Acrodermatitis chronica atrophicans (Europe) • Cardiac disease • Arthritis • Keratitis, uveitis • Neurologic disease

Although originally described as an oligoarticular arthritis, in recent years it has become evident that the significant involvement of Lyme disease in North America is neurologic (Table II) (6). Almost all levels of the neuraxis can be involved at various stages of this infection.

Lyme disease is our second major emerging infection after AIDS. More than 65,000 cases have been reported to the Centers for Disease Control and Prevention (CDC) since the disease became nationally reportable in 1982 (7). Lyme disease now accounts for 91% of vector-borne infections in this country (8). Although Lyme disease is geographically restricted to regions where the tick vector is endemic, studies have documented that this vector is spreading to new geographic areas, and that the tick infection rate within endemic regions is rising (9). Significant health care resources are being spent on the diagnosis and management of patients with presumed Lyme disease (10). Studies have documented that Lyme disease cases are both overdiagnosed (11,12) and underdiagnosed (13,14).

PROPERTIES OF THE BACTERIAL AGENT

B.burgdorferi belongs to the spirochete family (15). Spirochetes are technically gram negative bacteria. They are motile spiral-shaped bacteria up to 30 μm in length, with flagellae that lie between the inner and outer cell membranes. Spirochetes consist of three genera (*Borrelia,Leptospira,Treponema)*; four distinct spirochetal infections are recognized to cause disease in humans and to target the nervous system (Table III) (15,16). Except for syphilis, all the infections are zoonotic or vector-borne.

B.burgdorferi is only visualized under phase contrast or darkfield microscopy. Organisms are difficult to culture, and require special medium (Barbour-Stoenner-Kelly, BSK medium) to grow. There are two forms of BSK medium in current use. The older version, BSK-II, consists of bovine serum albumin, neopeptone, yeast extract, and

TABLE II

Neurologic Syndromes in North American Lyme Disease

Early characteristic syndromes	• "aseptic" meningitis,meningoencephalitis • cranial nerve palsy(esp uni or bilateral facial nerve palsy) • radiculoneuritis
Early unusual syndromes	• headache and stiff neck with normal CSF • confusional state • transient ischemic attack-like episodes
Late characteristic syndromes	• encephalopathy • sensorimotor polyradiculoneuropathy • encephalomyelitis, meningoencephalomyelitis
Late unusual syndromes	• intracranial hypertension syndrome (children, adolescents) • vasculitis • psychiatric syndrome • myositis • persistent headache with multisystem complaints

TABLE III

Human Spirochetal Infections which involve the nervous system

Flexible helical bacteria with endoflagellum Genera • *Treponema* • *Leptospira* • *Borrelia*	Neurologic infections occur with • *Treponema pallidum* (syphilis) • *Leptospira interrogans* (leptospirosis) • *Borrelia recurrentis* and other species(relapsing fever) • *B.burgdorferi* (Lyme disease)	Shared properties include • Skin or mucous membrane entry sites • Early spirochetemia • Wide dissemination • Chronic infection with episodic clinical disease • Protean disease syndromes

rabbit serum added to commercial tissue culture medium CMRL-1066 (17). BSK-H is a recent modified medium which does not contain gelatin or agarose, and which uses different optimized proportions of the BSK-II ingredients (18). It is typical for spirochetes to take several weeks to grow to sufficient densities to be visualized. Spirochetes tend to grow in aggregates early on, but can also be visualized to swim free in the medium.

The organism divides by transverse fission every 8 to 12 hours in vitro. In vivo, division may be even slower. Spirochete ultrastructure consists of an outer surface membrane, a periplasmic space containing 7 to 11 flagella, a peptidoglycan cell wall, and a trilaminar cytoplasmic membrane (19). In addition to one linear chromosome of

950 kB (20,21), *B.burgdorferi* contains extrachromosomal double stranded DNA in the form of supercoiled circular plasmids and linear plasmids. Anywhere from 4 to 10 distinct plasmids have been observed in various *B.burgdorferi* isolates (22). Virulent strains of *B.burgdorferi* usually have a larger number of plasmids than avirulent strains.

The chromosomal and extrachromosomal DNA code for a number of proteins. More than 100 polypeptides are distinguishable in a protein analysis of *B.burgdorferi*. There are constant major proteins which are not unique to *B.burgdorferi*, such as the p41 flagellin (shared by other spirochetes) and the p58-60 heat shock protein. *B.burgdorferi* contains additional p66 and 68, and p71 and 73 heat shock proteins, all of which are also not unique to the organism. These constant but nonspecific proteins cause serologic cross-reactivity. There are also variable major proteins which are unique to *B.burgdorferi*. These include the plasmid-coded outer surface lipoproteins (Osp): the 31 kilodalton (kD) OspA; the 34 kD OspB; and the 21-24 kD OspC. Other Osp proteins include OspD, OspE, and OspF. *B.burgdorferi* apparently also contains lipopolysaccharides with mitogenic and pyrogenic properties. Other important proteins are a unique p39 lipoprotein, and a unique p93 protein which appears to be a protoplasmic cylinder antigen. Isolates of *B.burgdorferi* show heterogeneity with regard to their DNA, plasmids, and variable proteins (23-27). The genospecies complex of *B.burgdorferi* sensu lato has been divided into more than 4 distinct genospecies: *B.burgdorferi* sensu stricto (most North American isolates), *Borrelia garinii, Borrelia afzelii, Borrelia japonica*, and other additional strains which do not fit into any of the specified genospecies (28-30). Ixodes ticks can be infected simultaneously with multiple strains of *B.burgdorferi*. Preliminary data suggests that different species cause different clinical disease patterns.

B.burgdorferi is believed to be an extracellular organism. Spirochetes readily adhere to and penetrate endothelial cells. In fact they can adhere to a number of different cell types, including glial cells, with attachment mediated by surface-exposed proteoglycans (31). In vitro *B.burgdorferi* invades fibroblasts to cause intracellular infection (32). The issue of whether *B.burgdorferi* in vivo causes intracellular infection is an important one that has therapeutic implications. Although numbers are small, there have been documented late isolations of *B.burgdorferi* from skin, cerebrospinal fluid (CSF), synovial tissue, flexor retinaculum, eye, myocardium, blood, and synovial fluid weeks to years after initial infection despite evidence for a host immune response (33). The ability of viable spirochetes to persist inside a cell, protected from extracellular antibiotics, is directly pertinent to the question of chronic infection.

CURRENT DIAGNOSTIC TECHNIQUES

The diagnosis of Lyme disease is ultimately a clinical one. Laboratory tests, however, are very useful to support the diagnosis (Table IV). With the exception of erythema migrans, the expanding red skin lesion which constitutes local infection, none of the clinical manifestations of Lyme disease are considered so characteristic as to be pathognomonic markers for *B.burgdorferi* infection. Therefore laboratory support for Lyme disease is very helpful to increase the likelihood of a correct diagnosis. In addition to body fluids such as blood, CSF, and synovial fluid, biopsy tissues (such as skin, cardiac muscle, synovium) are sometimes subjected to laboratory testing for Lyme disease.

Culture

Culture is considered the gold standard to document an infectious disease. Unfortunately, *B.burgdorferi* is difficult to culture or to stain in human body fluids and organs. The one exception is the early skin lesion. Skin punch biopsy of the erythema migrans lesion documents spirochetes in 60 to 86% of cases (34), although in a recent study lavage culture of skin lesions had a yield of only 29% (35). Other specimen sources are occasionally rewarding. At Stony Brook there have been six culture-positive CSF specimens in the past two years. All six cases had Lyme meningitis, an early dissemination neurologic syndrome. For the culture protocol 0.5cc of CSF was placed into 4.5cc of BSK-H medium (Sigma Chemical Co., St. Louis, MO), and then cultured at 33°C for up to 12 weeks. CSF was examined for spirochetes at several week intervals, using dark field microscopy. Spirochetes typically took at least 6 weeks to grow to detectable levels in CSF. Polymerase chain reaction (PCR) of cultured spirochetes confirmed that the organisms were *B.burgdorferi*, and limited heterogeneity suggested that there might be specific neurotropic spirochetal strains.

Serology

Detection of antibodies to *B.burgdorferi* is currently the most commonly used laboratory test to support a clinical diagnosis of Lyme disease. Unfortunately, there are a number of problems with current serologic testing for Lyme disease (Table V). Antibody assays are not standardized with regard to the antigens and reagents used, how the assays are run, the interpretation of positive and negative results, and the use of controls. Various laboratories have shown poor inter and intra assay reliability (36-39). Both false positive and false negative results can occur. Early on, before

TABLE IV

Current Laboratory Diagnosis of Lyme Disease

- Blood studies
  - -anti-*B.burgdorferi* antibodies (ELISA, Western blot)
  - -non-specific tests
  - -tests to exclude other diagnoses
  - -cell proliferation studies

- CSF studies
  - -intrathecal anti-*B.burgdorferi* antibody production
  - -cell count
  - -protein
  - -glucose
  - -VDRL
  - -other tests (oligoclonal bands, IgG index, myelin basic protein, etc.)
  - -experimental tests (antigen, PCR, culture, *Borrelial* immune complexes, specific IgM)

- Ancillary studies
  - -neuroimaging (MRI preferred)
  - -neurophysiologic (nerve conductions), evoked potentials, EMGs)
  - -neuropsychologic testing
  - -angiography or MR angiography
  - -EKG
  - -synovial fluid studies
  - -tissue biopsy (culture, PCR, histology)

TABLE V

Causes of Inccurate Lyme Serology (anti-*B.burgdorferi* antibodies)

False positives	False negatives
• other infections (spirochetal, bacterial, rickettsial, HIV-1,EBV) • hypergammaglobulinemia (polyclonal B cell activation) • diseases with high autoantibody levels • assay variability	• early infection • early antibiotics with blunted humoral response • assay variability

detectable levels of antibodies are synthesized, infected patients are seronegative. Some infected patients who receive early antibiotics (including amoxicillin, azithromycin, or doxycycline) never go on to synthesize detectable levels of free antibodies. A proportion of these seronegative patients, but certainly not all, will have detectable anti-*B.burgderferi* antibodies within circulating immune complexes (40,41). Because many of the immunogenic antigens of *B.burgdorferi* are shared with other organisms, legitimate false positive serologies occur.

The most commonly used serologic assays are enzyme linked immunosorbent (ELISA) assay, which have largely replaced indirect immunofluorescent (IFA) assays. Current ELISA assays utilize whole cells (tissue culture-grown spirochetes, which are washed and then sonicated), or purified protein extracts of these whole cells. Positive titers are often based on comparison to a negative control panel, for example optical density readings which are three standard deviations above the mean reading of the controls. Although their sensitivity is reasonable, specificity is a problem for the reasons noted above. Western immunoblot assays allow one to examine the specific reactivity of antibodies. The number of bands visualised on an IgM or IgG Western blot directly relates to Lyme disease duration and to disseminated, as opposed to local, disease (42). Recently the CDC and the Association of State and Territorial Public Health Laboratory Directors (ASTPHLD) recommended specific criteria for determining a positive IgM and IgG Lyme Western blot (Table VI) (43). These criteria may turn out to be overly restrictive. There are culture-confirmed Lyme disease cases who never develop a positive Western blot by these criteria (42).

One of the central difficulties in antibody testing for Lyme disease is that a positive serology documents prior exposure to the agent, but not active infection. However, intrathecal or intrasynovial compartmentalized organism-specific antibody production does provide indirect laboratory evidence for active infection. In this instance, preferential antibody synthesis is seen in CSF or synovial fluid, compared to the systemic blood compartment. With rare exception, in an untreated patient this is evidence for a central nervous system or joint infection by *B.burgdorferi*. Paired serum and CSF, or synovial fluid, samples are examined for anti-*B.burgdorferi* antibodies. There are several different techniques to determine intracompartmental antibody synthesis (44). These include performing antibody capture ELISA, and comparing the sample reading; normalizing paired samples by IgG concentration, and then running them at the same time in an indirect ELISA; or using serial sample dilutions to measure anti-*B.burgdorferi* antibodies as well as total immunoglobulin level. In this last case, the ratio of specific to total immunoglobulin is compared in CSF or synovial fluid relative to the ratio in serum.

PCR

PCR is being used to detect *B.burgdorferi* nucleic acid in a variety of body fluids as well as tissue samples. PCR has been applied to serum, CSF, synovial fluid, urine, blood, and plasma (45-48). Positivity rates have a wide range, depending on the reporting laboratory and investigator. PCR is most likely to be positive in skin lesions (70 to 90%), has a lower positivity rate for CSF, synovial fluid, and urine (20-40%), and has the lowest positivity rate for blood (9 to 26%). Both genomic (flagellin, 16S ribosomal RNA) and plasmid (OspA,OspB) primers have been used. Plasmid primers

TABLE VI

Recommendations of the CDC/ASTPHLD for serological diagnosis of Lyme disease
Two-test approach
Initial screening enzyme immunoassay (or immunofluorescent) assay
Confirm borderline or positive tests by Western immunoblot (order immunoglobulins G and M for illness<1 month, order immunoglobulin G only for illness>1 month)
Immunoglobulin M imunoblot (positive if two of three bands present)
P21-24 (outer surface protein C)
P39
P41 (flagellin)
Immunoglobulin G immunoblot (positive if five of 10 bands present)
P18
P21
P28
P30
P39
P41
P45
P58
P66
P93

appear to give a higher yield in clinical samples than genomic primers (49). The level of resolution of PCR can approach detection of a single spirochete in a clinical specimen. PCR assays for Lyme disease have not yet been standardized to the extent that they are in routine use. Protocols and primers optimized for sensitivity and specificity have not been determined. Although PCR can be extremely sensitive in infectious diseases, PCR applied to CSF in neurologic Lyme disease cases has shown sensitivity well below 100%. It is also recognized that false positive test results can occur with minimal contamination. This requires that the performing laboratory must be scrupulous in techniques to avoid low level contamination.

PCR detection of *B.burgdorferi* DNA does not indicate viable organisms. However, studies to date suggest that PCR positivity clears with effective antibiotic treatment (50).

TECHNIQUES IN DEVELOPMENT

Improved serologic assays

ELISA assays are more cost effective, efficient, and labor intensive for diagnostic screening purposes than Western blot assays. Subunit, whole, and chimeric

TABLE VII

CSF IgM Reactivity to *B.burgdorferi* in Early Neurologic Lyme Disease (LD)

Assay Reactivity	Early LD (N=16)				Chronic LD (N=13)	OND (N=19)
	EM 0-3wks (N=7)	EM 3-6wks (N=2)	EM 6-12wks (N=5)	disseminated (N=2)		
recombinant OspA (indirect ELISA)	3(43%)	1(50%)	1(20%)	2(100%)	3(23%)	0
recombinant OspB (indirect ELISA)	3(43%)	1(50%)	1(20%)	2(100%)	0	0
recombinant OspC (indirect ELISA)	4(57%)	1(50%)	5(100%)	2(100%)	3(23%)	0
whole cell B. burgdorferi proteins (indirect ELISA)	3/6 (50%)	1(50%)	4(80%)	2(100%)	10(77%)	0
whole cell B. burgdorferi proteins (antibody capture)	3(43%)	1(50%)	2(40%)	2(100%)	3/12(25%)	0/12

EM = erythema mygrans OND = other neurologic diseases

recombinant protein-based serologic assays are being developed for Lyme disease. Preliminary studies indicate that recombinant based ELISA assays will provide more accurate serologic test results than currently available tests (51,52). It is likely that a mixture of antigens (for example, in the form of specifically designed chimeric proteins) will provide a superior rapid screening serologic assay. Similarly, recombinant protein based Western blot assays may provide a more definitive way to document antibody specificity than the current molecular weight-based system.

OspC is immunodominant in the early immune response to *B.burgdorferi*, and IgM reactivity to recombinant OspC appears to be a better serologic marker for early infection than Western blot or ELISA using whole cell sonicate (53-55). In a preliminary study of early Lyme disease patients with neurologic involvement, including erythema migrans patients with headache, CSF anti-OspC IgM was detected in the majority of patients (Table VII). This IgM reactivity was more frequent than IgM to whole cell proteins.

<u>Antigen assays</u>

There is in vitro and in vivo evidence that *B.burgdorferi* releases pieces of its outer surface membrane in the form of "blebs" (56-61). These blebs have been

TABLE VIII

CSF OspA Antigen Capture ELISA

- Reagents: Mab 184.1, IgG2b, recognized epitope AA61; PAb $F(Ab')_2$, rabbit IgG, recognizes B31 strain whole cell preparation
- Detection limits: 0.03 nanograms
- Specificity: 99.3% (4/520 OND, including 37 CNS infection control)
- Sensitivity:
  - 100% Lyme radiculoneuritis
  - 60% Lyme meningitis
  - 47% Lyme facial nerve palsy
  - 36% multifocal EM with headache
  - 24% EM with headache
- Confirmatory tests:
  - 6/6 culture +
  - 7/7 PCR +
  - ~50% Western blot +

detected by electron microscopy, and contain antigens as well as DNA. There are a number of reports of *B.burgdorferi* antigen detection in body fluids of infected humans and animals (62-65). None of these tests are yet standardized for widespread use. For the past several years, the Stony Brook Department of Neurology research laboratory has been using an antigen capture ELISA for OspA detection in CSF, with good results (Table VIII) (66,67).

SUMMARY

There are two major diagnostic issues in Lyme disease which have yet to be resolved. One issue is the lack of an active infection assay. The second issue is the lack of a reliable antibody screening assay. Techniques are currently being developed to address the problem of active infection. These techniques involve standardization of PCR assays to detect *B.burgdorferi* DNA, standardization of antigen assays to detect *B.burgdorferi* proteins, and improved antibody assays to document infection of sequestered body compartments.

Techniques are also currently being developed to address the problem of sensitive and specific antibody assays. These techniques center around the use of critical immunodominant and specific recombinant *B.burgdorferi* proteins as antigen sources for both ELISA and Western blot assays.

REFERENCES

1. W Burgdorfer, AG Barbour, SF Hayes, et al. Science, 216, 1317-1319, (1982).
2. JL Benach, EM Bosler, JP Hanrahan, et al. N Engl J Med 308, 740-742 (1983).
3. AC Steere, RL Grodzicki, AN Kornblatt et al. N Engl J Med 308, 733-740 (1983).
4. JP Hanrahan, JL Benach, JL Coleman, et al. J Infect Dis, 150, 489-496 (1984).
5. AC Steere, E Taylor, ML Wilson, et al. J Infect Dis, 154, 295-300 (1986).
6. JC Garcia-Monco and JL Benach. Ann Neurol, 37, 691-702 (1995).
7. Centers for Disease Control and Prevention. MMWR, 44(24), 459-462 (1994).
8. DT Dennis in: Lyme Disease, PK Coyle, ed, Mosby Year Book, St. Louis, (1993) pp. 27-37.
9. DJ White, H-G Chang, JL Benach et al. JAMA, 266, 1230-1236 (1991).
10. Centers for Disease Control and Prevention/DVBID, 4, 2-3 (1993).
11. AC Steere, E. Taylor, GL McHugh and EL Logigian. JAMA 269, 1812-1816 (1993).
12. CD Rose, PT Fawcett, KM Gibney, and RA Doughty. Clin. Pediatr. 33, 663-668 (1994).
13. Centers for Disease Control and Prevention. MMWR 42, 348-350 (1993).
14. BS Coyle, GT Strickland, YY Liang, et al. J Infect Dis, 173, 1260-1262 (1996).
15. GP Schmid. Rev Infect Dis, 11, S1460-1469 (1989).
16. SA Lukehart, EW Hook III, SA Baker-Zander, et al. Ann Intern Med 109, 855-862 (1988).
17. Barbour AG. Yale J Biol Med, 57, 521-525 (1984).
18. Pollack RJ, Telford SR III, Spielman A. J Clin Microbiol 31, 1251-1255 (1993).
19. PA Rosa and TG Schwan, in: Lyme Disease, PK Coyle, ed, Mosby Year Book, St. Louis, (1993) pp. 8-17.
20. C Baril, C Richaud, G Baranton, and IS Girons. Res. Microbiol. 140, 507-516 (1989).
21. MS Ferdows and AG Barbour. Proc. Natl. Acad. Sci. USA 86, 5969-5973 (1989).
22. Y Xu and RC Johnson. J Clin. Microbiol. 33, 2679-2685 (1995).
23. AG Barbour. Scand J Infect Dis, S77, 88-93, (1991).
24. B Wilske, JF Anderson, G Barcanton, et al. Scand J Infect Dis, S77, 108-129 (1991).
25. U Busch, C Hizo-Teufel, R Boehmer, et al. J Clin. Microbiol. 34, 1072-1078 (1996).
26. D Liveris, GP Wormser, J Nowakowski, et al. J Clin Microbiol. 34, 1306-1309 (1996).
27. RN Picken, Y Cheng, D Han, et al. J Clin Microbiol. 33, 2304-2315 (1995).
28. G Baranton, D Postic, IS Girons, et al. Int J Syst Bacteriol. 42, 378-383 (1992).
29. DS Guttman, PW Wang, I-N Wang, et al. J Clin. Microbiol. 34, 652-656 (1996).
30. Y Yanagihara, H Kawabata and T Masuzawa, in Advances in Lyme borreliosis research: proceedings of the VI International Conference on Lyme Borreliosis, R Cercini, V Sambri and M LaPlaca, eds. Societa Editrice Esculapo, Bologna, Italy, (1994) pp. 27-30.
31. RD Isaacs. J Clin Invest, 93, 809-819 (1994).
32. K Georgilis, M Peacocke and MS Klempner. J Infect Dis 166, 440-444 (1992).
33. MS Klempner, R Noring and RA Rogers. J Infect Dis, 167, 1074-1081, (1993).
34. BW Berger, RC Johnson, C Kodner and L Coleman. J Clin Microbiol, 30, 359-361 (1992).
35. GP Wormser, G Forseter, D Cooper, et al. JAMA, 268, 1311-1313 (1992).
36. M Corpuz, E Hilton, MP Lardis, et al. Arch Intern Med 151, 1837-1840 (1991).
37. CW Hedberg, MT Osterholm, KL MacDonald, et al. J Infect Dis, 155, 1325-1327 (1987).

38. SW Luger and E Krauss. Arch Intern Med, 150, 761-763 (1990).
39. BS Schwartz, MD Goldstein, JMC Ribeiro, et al. JAMA 262, 3431-3434 (1989).
40. SE Schutzer, PK Coyle, AL Belman, et al. Lancet 1, 312-315 (1990).
41. SE Schutzer, PK Coyle, JJ Dunn, et al. J Clin Invest 94, 454-457 (1994).
42. ME Aguero-Rosenfeld, J Nowakowski, S Bittker et al. J Clin Microbiol. 34. 1-9 (1996).
43. Centers for Disease Control. MMWR 44, 590-591 (1995).
44. PK Coyle, in Lyme Disease, Mosby Year Book. St. Louis, (1993) pp.136- 148.
45. JF Bradley, RC Johnson and JL Goodman. Ann Intern Med. 120, 487-489 (1994).
46. H Karch, HI Huppertz, M Bohme, et al. J Clin Microbiol. 32, 2312-2314 (1994).
47. BJ Luft, CR Steinman, HC Neimark, et al. JAMA, 267 , 1364-1367 (1992).
48. SE Moter, H Hofmann, R Wallich, et al. J Clin Microbiol. 32, 2980-2988 (1994).
49. DH Persing, BJ Rutledge, PN Rys, et al. J Infect Dis 169, 668-672 (1994).
50. SE Malawista, SW Barthold, DH Persing. J Infect Dis 170, 1312-1316 (1994).
51. LA Magnarelli, E Fikrig, SJ Padula, et al. J Clin Microbiol. 34, 237-240 (1996).
52. S Rauer, M Kayser, U Neubert, et al. J Clin Microbiol. 33, 2596-2600 (1995).
53. BP Fung, GL McHugh, JM Leong, and AC Steere. Infect Immun. 62, 3213-3221 (1994).
54. MA Gerber, ED Shapiro, GL Bell, et al. J Infect Dis. 171, 724-727 (1995).
55. SJ Padula, F Dias, A Sampiere, et al. J Clin Microbiol. 32, 1733-1738 (1994).
56. AG Barbour and SF Hayes. Microbiol Rev, 50, 381-400 (1986).
57. V Brade, I Kleber, G Acker. Immunobiol, 185, 453-465 (1992).
58. DW Dorward, TG Schwan, and CF Garon. J Clin Microbiol 289, 1162-1170 (1991).
59. CF Garon, DW Dorward and MD Corwin. Scanning Microsc Suppl 3, 109-115 (1989).
60. LI Katona, G Beck and GS Habicht. Infect Immun, 60, 4995-5003 (1992).
61. WM Whitmire and CF Garon. Infect Immun, 61, 1460-1467 (1993).
62. JL Benach, JL Coleman and MG Golightly. J Immunol, 140, 265-272 (1988).
63. DW Dorward, TG Schwan and CF Garon. J Clin Microbiol, 29, 1162-1170 (1991).
64. JC Garcia-Monco, BF Villar, JC Alen and JL Benach. J Infect Dis, 161, 1187-1193 (1990).
65. FW Hyde, RC Johnson, TJ White and CE Shelburne. J Clin Microbiol 27, 58-61 (1989).
66. PK Coyle, Z Deng, SE Schutzer, et al. Neurol, 43, 1093-1097 (1993).
67. PK Coyle, SE Schutzer, Z Deng, et al. Neurol 45, 2010-2015 (1995).

THE TOXINS OF GROUP A STREPTOCOCCUS, THE FLESH EATING BACTERIA

Dennis L. Stevens
Division of Infectious Diseases
Veterans Affairs Medical Center
Boise, Idaho

Department of Medicine
University of Washington School of Medicine
Seattle, Washington

ABSTRACT

Streptococcus pyogenes causes a wide variety of infections in individuals of all ages in most countries of the world. Because of the frequency with which these infections occur, physicians are quite familiar with the diversity of clinical presentations associated with the Group A streptococcus. Yet in the late 1980's, a severe form of streptococcal infection, the Streptococcal Toxic Shock Syndrome, emerged and has persisted for the last 10 years. This syndrome is associated with invasive soft tissue infections and the early onset of shock and organ failure. The purpose of this paper is to briefly describe the epidemiologic and clinical features of the Streptococcal Toxic Shock Syndromes and to emphasize the role that toxins produced by *S. pyogenes* play in the pathogenesis of this disease.

INTRODUCTION

The epidemiology of group A streptococcal infections is no less complicated than its virulence factors. Epidemics of scarlet fever have been well documented since the 16th century and were commonly associated with high mortality. In contrast, the incidence of severe forms of scarlet fever has greatly diminished throughout most of the 20th century - a decline which began well before the development of penicillin (1). Similarly, rheumatic fever has declined in frequency and severity since World War II, due, in part, to the availability of penicillin. These declines have been very dramatic in the western world, though both scarlet fever and rheumatic fever continue to be significant, but poorly defined, problems in underdeveloped countries. These temporal and geographical differences in incidence and severity of disease in developed countries suggest that either the virulence of group A streptococcus has waned or, alternatively, the resistance of the human host to streptococcal infections has increased during the present century.

Streptococcal Toxic Shock Syndrome

A. The Emergence of Streptococcal Toxic Shock Syndrome (Strep TSS).

In the latter half of the 1980's, western countries began to experience a re-emergence of rheumatic fever (2) and later fulminant group A streptococcal infections associated with shock, organ failure, necrotizing fasciitis, bacteremia and death termed the Streptococcal Toxic Shock Syndrome (Strep TSS; reviewed in (3)). Since the first descriptions of Strep TSS in the United States (4-7), over 100 additional reports have been published from the United States, Europe, Canada, Australia, New Zealand and the Orient (reviewed in (3)). Cases tend to be sporadic with a prevalence of no more than 10-20 cases per 100,000 population (8-10). Secondary cases are rare but have been reported among family members where intimate contact with a case has occurred (11,12), in nursing homes settings (11,13-15) and among medical personnel caring for patients with Strep TSS (11,14,16).

It is clear that a wide variety of primary group A streptococcal infections may result in Strep TSS. In addition, not all the reported features of Strep TSS occur in every patient and finally some patients with group A streptococcal infections may develop shock or death, but only late in the course of illness. The spirit of the Working Group on Severe Group A Streptococcal Infections (17) was to provide a case definition of Strep TSS which required that shock and organ failure occur early in the course of infection In addition, this definition acknowledges that the sites of infection may be diverse, and not limited to necrotizing fasciitis, though the latter has been commonly associated with Strep TSS (3).

B. Symptoms of Strep TSS

Approximately 20% of patients experience an influenza-like syndrome characterized by fever, chills, and myalgias (7,18). Other patients may experience nausea, vomiting, and diarrhea associated with fever prior to development of Strep TSS. Pain, the most common initial symptom of Strep TSS, and is abrupt and severe

(7,18) and commonly precedes tenderness or any physical evidence of localized infection. It is usually localized to an extremity, though headache or pain mimicking sinusitis, peritonitis, pelvic inflammatory disease, acute myocardial infarction or pericarditis have also been described (7,18). The presence of fever together with increasingly severe pain should alert the physician to the possibility of a deep seated *S. pyogenes* infection.

C. Physical Findings of Strep TSS

Fever, the most common presenting sign, may range from 99-106 degrees F, except in patients with advanced disease who may develop profound hypothermia secondary to shock (7,18). Confusion is present in 55% of patients and, in some, coma or combativeness are manifest (7,18). Evidence of soft tissue infection such as localized swelling, tenderness, pain and erythema are often present at the time of admission and are apparent early where a cutaneous portal of entry is defined. The appearance of violaceous bullae suggest deeper soft tissue infection such as necrotizing fasciitis or myositis. In one series, 70% of patients required surgical debridement, fasciotomy or amputation (18). The remainder of these cases have displayed a variety of clinical presentations such as endophthalmitis, myositis, perihepatitis, peritonitis, myocarditis, puerperal sepsis, septic arthritis, pharyngitis and overwhelming sepsis (Table I) (7,18,19).

Hypotension and tachycardia out of proportion to the fever are present in all patients at some time in their course, though 40% of patients in one series had normal blood pressure at the time of initial evaluation (7,18). Massive local swelling is apparent in patients with

TABLE I

Streptococcal Infections Associated with Strep TSS
1. necrotizing fasciitis
2. myonecrosis
3. pneumonia
4. septic joint
5. peritonitis
6. sinusitis
7. epiglottitis
8. meningitis
9. cellulitis
10. pharyngitis (rare)

necrotizing fasciitis and/or myositis, however, generalized interstitial edema develops in virtually all patients due to diffuse capillary leakiness.

D. Laboratory Test Results in Patients with the Strep TSS

Though the initial white blood cell count may be normal or only slightly elevated, the differential count demonstrates a strikingly high percentage of immature neutrophils including band forms, metamyelocytes, and myelocytes; the mean percentage exceeds 50% (7,18). Evidence of renal involvement is apparent in 80% of patients (mean serum creatinine > 2.5 times normal) (7,18). Hypoalbuminemia and profound hypocalcemia are prominent features at the time of admission and throughout the hospital course. The serum creatinine phospho-kinase level may be useful in detecting the presence of deeper soft-tissue infections, and when the level was elevated or rising, there was a good correlation with necrotizing fasciitis or myositis (7,18).

E. Clinical Course

Hypotension was apparent at the time of admission or within 4 - 8 hours in virtually all patients (Table II). In 10% of patients, systolic blood pressure normalized promptly hours after administration of antibiotics, fluid resuscitation or infusion of dopamine. In the remaining patients, shock and renal dysfunction progressed or persisted for 48 - 72 hours in spite of treatment, and several patients required dialysis for 10 - 20 days (7,18). Among survivors, renal function returned to normal within 4 - 6 weeks. Renal dysfunction preceded shock in many cases and was apparent early in the course of shock in all others. Adult respiratory distress syndrome (ARDS) occurs most commonly after hypotension and may be found in 55% of patients(7). Sixty percent of patients with Strep TSS were bacteremic and overall mortality has ranged from 30% (7,18) to 60 - 70% (9,10,20). Morbidity has also been high and approximately 50% of patients required major surgical procedures, which included fasciotomy, surgical debridement, exploratory laparotomy, intraocular aspiration, amputation and hysterectomy (7,18).

F. Predisposing conditions, portals of entry and factors affecting host response to infection.

Despite the fact that the strains of group A streptococcus associated with Strep TSS are found widely among asymptomatic

TABLE II

Complications of Group A Streptococcal Soft-Tissue Infection

COMPLICATION	PERCENTAGE OF PATIENTS
Shock	95
ARDS	55
Renal impairment	80
Irreversible	10
Reversible	70
Bacteremia	60
Mortality	30

individuals in communities, cases of Strep TSS have remained sporadic and epidemics have not materialized, suggesting that simple contact with a virulent strain is not sufficient to cause Strep TSS. Thus, other predisposing factors may be necessary for Strep TSS to occur. For example, certain viral infections such as influenza and chicken pox can disrupt anatomical barriers such as the mucosa and skin. Similarly, influenza virus infection alters respiratory epithelium sufficiently to provide a portal of entry. As stated previously influenza and chicken pox are associated with Strep TSS and could predispose humans to develop worse outcomes based upon deleterious effects upon the immune system. Disruption of anatomical barriers from other causes such as lacerations, burns, slivers, surgical procedures, decubitus ulcers, intravenous drug abuse, bites (insects, dogs, cats etc.) provide cutaneous portals of entry.

Childbirth, another risk factor, compromises the integrity of the uterine mucosa, allowing entry of the organism. The source of streptococci in this setting is usually group A streptococcus which are colonizing the vaginal mucosa, though in the past bacteria contamination of the vagina and birth canal from hospital environment or personnel (21) was well described.

In some series of Strep TSS, a portal of entry cannot be ascertained (7,18,22), and 50% of patients may not have a distinct diagnosis, until shock and organ failure develop. In these cases, very specific predisposing factors are likely required. For example, these deep soft tissue infections invariably occur at the exact site of blunt trauma, muscle strain, hematoma or joint effusion. These observations raise two important questions. First, how does non-penetrating local trauma predispose to infection developing at that exact site? And second, in the absence of a penetrating injury, how does group A streptococcus arrive at the site of infection? The best hypothesis is that the organism translocates to the site of injury likely through a transient bacteremia from a mucous membrane, usually the throat. A diagnosis may be difficult at this time because the infection is so deep in muscle or fascia that there may not be evidence of cellulitis, lymphangitis, or erysipelas. This confusing constellation of signs and symptoms results in delays in diagnosis or an incorrect diagnosis such as thrombophlebitis or merely musculoskeletal strain.

G. Virulence Factors and Pathogenesis

The group A streptococcus' evolution with the human host has favored its development of adaptative ways to interact and even inactivate effector molecules produced by the human immune system.

This section will discuss these virulence factors in the context of their role in pathogenesis. Pyrogenic exotoxins induce fever in humans and animals and also participate in shock by lowering the threshold to exogenous endotoxin (18). Streptococcal pyrogenic exotoxins A (SPEA) and B (SPEB) induce human mononuclear cells to synthesize not only tumor necrosis factor- α (TNFα) (23) but also interleukin-1β (IL-1β) (24) and interleukin-6 (IL-6) (24) suggesting that TNF could mediate the fever, shock and tissue injury observed in patients with Strep TSS (7). Interestingly, fever induced by injection of purified SPEA into rabbits is totally abolished by prior administration of a neutralizing monoclonal antibody against rabbit TNF (author's unpublished observations). Pyrogenic exotoxin C (SPEC) has been associated with mild cases of scarlet fever in the United States (author's observations) and in England (25). The roles of two newly described pyrogenic exotoxins, streptococcal superantigen (SSA) (26) and mitogenic factor (MF) (27,28), in Strep TSS have not been elucidated, though both are capable of inducing synthesis of a variety of cytokines.

M-protein contributes to invasiveness through its ability to impede phagocytosis of streptococci by human polymorphonuclear leukocytes (PMNL) (29). Conversely, type specific antibody against the M-protein enhances phagocytosis (29). Following infection with a particular M-type, specific antibody confers resistance to challenge to viable group A streptococcus of that M-type (29). In addition, M-proteins and certain M-like proteins have the ability to bind various classes of immunoglobulins including IgG and IgM. While M types 1 and 3 have been the most common strains isolated from cases of Strep TSS, many other M types, including M-12, M-28, M-6, M-18 and some non-typable strains, have also been isolated from such cases. M types 1 and 3 are also commonly isolated from asymptomatic carriers, patients with pharyngitis, and mild scarlet fever (30,31). Using multi-locus electrophoresis, Musser (32) has shown that strains from patients with Strep TSS fit one of two genetic patterns, termed ET-1 or ET-2, which correspond to M-1 and M-3 strains, respectively. Cleary (33) studied M-1 strains from invasive cases and non-invasive cases and was able to demonstrate two distinct DNA fingerprint patterns that he referred to invasin positive or invasin negative, respectively. The invasin positive strains contained additional genetic elements, at least one of which coded for SPEA.

The mere presence of a potential virulence factor by such strains may be less important than the quantity that is produced. Quantitation of putative virulence factors from strains of group A streptococcus isolated from patients with Strep TSS has not been definitive, but Cleary (34) has proposed that the organism possess a virulence regulon that controls the production of a gene cluster of virulence factors such as M-protein and C5a- peptidase. Strains of group A streptococcus may acquire genetic information coding for SPEA or SPEC via bacteriophage and following lysogenic conversion, toxin is synthesized synchronously with growth cycle of the streptococcus (35-37).

It has been suggested that streptococcal and staphylococcal toxic shock syndromes could be related to the ability of certain streptococcal and staphylococcal factors to act as "superantigens" (38). There is data to suggest that SPEA, SPEB, SPEC, MF, SSA and M-protein as well as staphylococcal TSST-1 and enterotoxins A, B, and C can stimulate T cell responses through their ability to bind to both the Class II MHC complex of antigen presenting cells and the Vβ region of the T cell receptor (38). The net effect would be to induce T cell stimulation with production of cytokines capable of mediating shock and tissue injury. Recently, Hackett and Stevens demonstrated that SPEA induced both TNFα and TNFβ from mixed cultures of monocytes and lymphocytes (39), supporting the role of these lymphocyte-derived cytokines in shock associated with strains producing SPEA. Dale et al (40) demonstrated that purified polypeptide fragments of pepsin extracted M proteins induced blastogenic responses in human lymphocytes but not laboratory animals. Kotb subsequently (41) has shown that a digest of M-protein type 5 (pep M5) can also stimulate T cell responses by this superantigen mechanism. Induction of cytokines by monocytes was dependent upon the presence of T-lymphocytes (41). Interestingly, the requirement for intact T-cells could be overcome by the addition of exogenous interferon gamma (41). Pep M5 also stimulated mixed cultures of monocytes and lymphocytes to produce the lymphokines, interferon gamma and TNFβ, though TNFβ production was greatly enhanced by the addition of exogenous gamma interferon (41). These in vitro results suggest that superantigens could play a role in the pathogenesis of Strep TSS. Proof would require demonstration of massive expansion of T-cell subsets bearing Vβ repertoires specific for the putative superantigen(s). Recently, quantitation of such T-cell subsets in patients with acute Strep TSS demonstrated deletion rather than expansion, suggesting that perhaps the life-span of the expanded subset was shortened by a process of apoptosis (42). In addition, the subsets deleted were not specific for SPEA, SPEB, SPEC, or MF suggesting that perhaps an as yet undefined superantigen may play a role (42).

Cytokine production by less exotic mechanisms may also contribute to the genesis of shock and organ failure. Peptidoglycan, lipoteichoic acid (43) and killed streptococci (44,45) are capable of inducing TNFα production by mononuclear cells in vitro (18,45,46). Exotoxins such as streptolysin O (SLO) are also potent inducers of TNFα and IL-1β. SPEB, a proteinase precursor, has the ability to cleave pre-IL-1β to release preformed IL-1β (47). Finally, SLO and SPEA together have additive effects in the induction of IL-1β by human mononuclear cells (39). Whatever the mechanisms, induction of cytokines in vivo is likely the cause of shock, and many fluid phase streptococcal virulence factors as well as cell wall components are potent inducers of TNF and IL-1. Thus, the specific virulence factors that induce cytokine production in patients may be complex, particularly among the 60% of Strep TSS patients who are bacteremic (7). Here the systemic immune system would be exposed to a barrage of streptococcal virulence factors all of which are capable of cytokine induction through a variety of cellular mechanisms.

The interaction between these microbial virulence factors and an immune or non-immune host determines the epidemiology, clinical syndrome and outcome. Since horizontal transmission of group A streptococcus is well documented, the only explanation for the absence of a high attack rate of invasive infection is the presence of significant herd immunity against one or more of the virulence factors responsible for Strep TSS (10). This model explains why (1) epidemics have not materialized and (2) why a particular strain of group A streptococcus can cause different clinical manifestations in the same community (48).

G. Treatment of Streptococcal Toxic Shock Syndrome

S. pyogenes remains exquisitely susceptible to ß-lactam antibiotics, and numerous studies have demonstrated the clinical efficacy of penicillin for treating erysipelas, impetigo, pharyngitis and cellulitis. Nonetheless, clinical failures of penicillin treatment of streptococcal infection do occur. Specifically it is the more aggressive group A streptococcal infections (such as necrotizing fasciitis, empyema, burn wound sepsis, subcutaneous gangrene, and myositis) that respond less well to penicillin and these continue to be associated with high mortality and extensive morbidity (7,18,49-53). For example, a recent report of 25 cases of streptococcal myositis reported an overall mortality of 85% in spite of penicillin therapy (49). Finally, several studies in experimental infection suggest that penicillin fails when large numbers of organisms are present (54,55). For example, in a mouse model of myositis due to *S. pyogenes*, penicillin was ineffective when treatment was delayed ≥2 hours after initiation of infection (55)]. Survival of erythromycin-treated mice was greater than that of both penicillin-treated mice and untreated controls, but only if treatment was begun within 2 hours. Mice receiving clindamycin, however, had survival rates of 100%, 100%, 80%, and 70% when treatment was delayed 0, 2, 6, and 16.5 hours, respectively (55,56).

Eagle suggested that penicillin failed in this type of infection because of the "physiologic state of the organism" (54). This phenomenon has been explained at least in part, by in vitro and in vivo inoculum effects (57,58). This in turn is related to the stage of growth of the bacteria. In vivo, early in the stages of infection or in mild infections organisms are growing rapidly, are present in rather small numbers and respond well to beta lactam antibiotics. With delays in treatment, higher concentrations of streptococci accumulate and growth begins to slow (54). Since penicillin mediates its anti-bacterial action against group A streptococcus by intimately interacting with the expressed penicillin binding proteins (PBP), we wondered whether the expression of PBP's was constant throughout the growth cycle. Therefore we compared the penicillin-binding protein patterns from membrane proteins of group A streptococci isolated from different stages of growth, i.e., mid-log-phase and stationary-phase. Binding of radiolabeled penicillin by all PBPs was decreased in stationary cells. In fact, PBPs 1 and 4 were undetectable at 36 hours (57). Thus, the loss of certain PBPs during stationary-phase growth in vitro may be responsible for the inoculum effect observed *in vivo* and may account for the failure of penicillin in both experimental and human cases of severe streptococcal infection.

The greater efficacy of clindamycin in severe group A streptococcal infections is due to many factors. First, its efficacy is not affected by inoculum size or stage of growth (57,59). Secondly, clindamycin is a potent suppressor of bacterial toxin synthesis (60,61). Third, clindamycin facilitates phagocytosis of *S. pyogenes* by inhibiting M-protein synthesis (61). Fourth, clindamycin suppresses synthesis of penicillin-binding proteins which, in addition to being targets for penicillin, are also enzymes involved in cell wall synthesis and degradation (59). Fifth, clindamycin has a longer post-antibiotic effect than ß-lactams such as penicillin. Lastly, we have recently shown that clindamycin causes suppression of LPS-induced monocyte synthesis of TNFα (62). Thus, clindamycin's efficacy may also be related to its ability to modulate the immune response.

H. Other Treatment Measures

Aggressive exploration and debridement of suspected deep seated *S. pyogenes* infection is mandatory. Exploration through a small incision with visualization of muscle and fascia, may be necessary to establish the depth and severity of infection and to obtain material for Gram stain and culture (19).

Because of intractable hypotension and diffuse capillary leak, massive amounts of intravenous fluids (10-20 liters/day) are often necessary. Pressors such as dopamine are frequently also required.

Neutralization of circulating toxins would be a desirable therapeutic modality yet specific antibody preparations are not commercially available in the United States or Europe. However, some commercial intravenous gamma globulin preparations (IVIG) do in fact contain neutralizing antibodies against several streptococcal factors such as the pyrogenic exotoxins (i.e., SPEA, SPEB, SPEC and MF) as well as SLO and DNase B (10,63). Recently, two case reports have been published which describe two patients who dramatically improved following administration of intravenous gamma globulin (64,65). In another publication, four patients received either intramuscular or intravascular gamma globulin preparations and 7 others received fresh frozen plasma (66). In the largest series, published in abstract form, 15 patients were given IVIG and mortality was reduced from 66% to 13% (67). None of these studies, were randomized and the latter one used historical controls for comparison.

Gamma globulin "non-specifically" inhibits TNF alpha production by either LPS stimulated monocytes (68) or splenocytes from animals with adjuvant induced arthritis (69). Thus, IGIV might be efficacious by specific neutralization of streptococcal toxins, by non-specific inhibition of monocyte/ T-cell activation or by inhibition of other putative streptococcal virulence factors. Recently a monoclonal antibody against TNFα showed promising efficacy in a baboon model of Strep TSS (70).

SUMMARY

In summary, if a wild "flesh eating strain" has recently emerged, a major epidemic with a high attack rate would be expected. Clearly epidemics of streptococcal infections, including impetigo, pharyngitis, scarlet fever and rheumatic fever, have occurred in the past. In the last decade subsequent to early reports of Strep TSS, we conclude that the incidence has remained relatively low. Large outbreaks have not occurred because (1) the vast majority of the population probably has immunity to one or more streptococcal virulence factors (18,66); (2) predisposing conditions (varicella, use of NSAIDs, etc.) are required in a given patient (71); and (3) there may be only a small percentage of the population having an inherent predisposition to severe streptococcal infection by virtue constitutional factors such as HLA Class II antigen type (72,73), B-cell allogantigens (74), or specific Vβ regions on lymphocytes. This hypothesis is further supported by the observation that secondary cases of Strep TSS, though reported (11), have been rare.

REFERENCES

1. T.M. Rotch, Pediatrics: The Hygienic and Medical Treatment of Children. J.B. Lippincott Company, Philadelphia, (1896)

2. L.G. Veasy, S.E. Wiedmeier and G.S. Orsmond. N Engl J Med, 316, 421-427 (1986).

3. D. Stevens. Emerg Infect Dis, 1, 69-78 (1995).

4. R.N. Greenberg, B.G. Willoughby, D.J. Kennedy, T.J. Otto, R. McMillian and T.G. Bloomster. Southern Medical Journal, 76, No. 7, 916-918 (1983).

5. L.A. Cone, D.R. Woodard, P.M. Schlievert and G.S. Tomory. N Engl J Med, 317, 146-149 (1987).

6. T. Bartter, A. Dascal, K. Carroll and F.J. Curley. Arch Intern Med, 148, 1421-1424 (1988).

7. D.L. Stevens, M.H. Tanner, J. Winship, R. Swarts, K.M. Reis, P.M. Schlievert and E. Kaplan. N Eng J Med, 321(1), 1-7 (1989).

8. C.W. Hoge, B. Schwartz, D.F. Talkington, R.F. Breiman, E.M. MacNeill and S.J. Englender. JAMA, 269, 384-389 (1993).

9. B. Demers, A.E. Simor, H. Vellend, P.M. Schlievert, S. Byrne, F. Jamieson, S. Walmsley and D.E. Low. Clin Infect Dis, 16, 792-800 (1993).

10. S.E. Holm, A. Norrby, A.M. Bergholm and M. Norgren. J Infect Dis, 166, 31-37 (1992).

11. B. Schwartz, J.A. Elliot, J.C. Butler, P.A. Simon, B.L. Jameson, G.E. Welch and R.R. Facklam. Clin Infect Dis, 15(2), 277-284 (1992).

12. J.R. Dipersio, L.A. Define, W. Gardner, D.L. Stevens, E.L. Kaplan and T.M. File. Program and Abstracts of the American Society for Microbiology, Abstract C189 (1995).

13. S.B. Auerbach, B. Schwartz, D. Williams, M.G. Fiorilli, A.A. Adimora, R.F. Breiman and W.R. Jarvis. Arch Intern Med, 152(5), 1017-1022 (1992).

14. J.J. Hohenboken, F. Anderson and E.L. Kaplan. ICAAC, Abstract# J189 (1994).

15. K. McGeer, R. Green and D. Cann. Program and Abstracts of the 35th Interscience Conference on Antimicrobial Agents and Chemotherapy, Abstract K135 (1995).

16. T.D. Valenzuela, T.M. Hooton, E.L. Kaplan and P.M. Schlievert. Ann Emerg Med, 20:1, 123-125 (1991).

17. The Working Group on Severe Streptococcal Infections. JAMA, 269(3), 390-391 (1993).

18. D.L. Stevens. Clin Infect Dis, 14, 2-13 (1992).

19. A.I. Bisno and D.L. Stevens. N Engl J Med, 334, 240-245 (1996).

20. J.E. Alouf and C. Geoffroy, in Sourcebook of Bacterial Protein Toxins J.E. Alouf and J.H. Freereds, Academic Press, New York, 1991) pp. 147-186.

21. W.E. Stamm, J.C. Feeley and R.R. Facklam. J Infect Dis, 138, 287-292 (1978).

22. M.R. Bibler and G.W. Rouan. Rev Infect Dis, 8, 941-951 (1986).

23. D.J. Fast, P.M. Schlievert and R.D. Nelson. Infect Immun, 57, 291-294 (1989).

24. S.P. Hackett, P.M. Schlievert and D.L. Stevens. Clin Res, 39, 189A (1991).

25. G. Hallas. J Hyg (Camb), 95, 47-57 (1985).

26. J.A. Mollick, G.G. Miller, J.M. Musser, R.G. Cook, D. Grossman and R.R. Rich. J Clin Invest, 92, 710-719 (1993).

27. M. Iwasaki, H. Igarashi, Y. Hinuma and T. Yutsudo. Febs Lett, 331, 187-192 (1993).

28. A. Norrby-Teglund, D. Newton, M. Kotb, S.E. Holm and M. Norgren. Infect Immun, 62, 5227-5233 (1994).

29. R.C. Lancefield. J Immunol, 89, 307-313 (1962).

30. D.R. Johnson, D.L. Stevens and E.L. Kaplan. J Infect Dis, 166, 374-382 (1992).

31. W. Kohler, D. Gerlach and H. Knoll. Zbl Bakt Hyg, 266, 104-115 (1987).

32. J.M. Musser, A.R. Hauser, M.H. Kim, P.M. Schlievert, K. Nelson and R.K. Selander. Proc Natl Acad Sci USA, 88, 2668-2672 (1991).

33. P.P. Cleary, E.L. Kaplan, J.P. Handley, A. Wlazlo, M.H. Kim, A.R. Hauser and P.M. Schlievert. Lancet, 339, 518-521 (1992).

34. R. Cleary, C. Chen, D. Lapenta, N. Bormann, D. Heath and E. Haanes. Program and Abstracts of the Third International ASM conference on Streptococcal Genetics, Abstract #19 (1990).

35. S.K. Nida and J.J. Ferretti. Infect Immun, 36(2), 745-750 (1982).

36. A.R. Hauser, S.C. Goshorn, E. Kaplan, D.L. Stevens and P.M. Schlievert. Third International ASM conference on Streptococcal Genetics, Minneapolis, MN (1990).

37. L.P. Johnson, M.A. Tomai and P.M. Schlievert. J Bacteriol, 166, 623-627 (1986).

38. J.A. Mollick and R.R. Rich. Clin Res, 39(2), 213A (1991).

39. S.P. Hackett and D.L. Stevens. J Infect Dis, 165, 879-885 (1992).

40. J.B. Dale, W. Simpson, I. Ofek and E.H. Beachey. J Immunol, 126, 1499-1505 (1981).

41. M. Kotb, M. Tomai, G. Majumdar, J. Walker and E.H. Beachey. XI Lancefield International Symposium on Streptococcal Diseases, Sienna, Italy (1990).

42. R. Watanabe-Ohnishi, D.E. Low, A. McGeer, D.L. Stevens, P.M. Schlievert, D. Newton, B. Schwartz, B. Kreiswirth, Ontario Streptococcal Study Project and M. Kotb. J Infect Dis, 171, 74-84 (1995).

43. D.L. Stevens, A.E. Bryant and S.P. Hackett. Curr Opin Infect Dis, 5(3), 355-363 (1992).

44. S. Hackett, J. Ferretti and D. Stevens. Program and Abstracts of the American Society for Microbiology, Abstract B-249 (1994).

45. H. Muller-Alouf, J.E. Alouf, D. Gerlach, J.H. Ozegowski, C. Fitting and J.M. Cavaillon. Infect Immun, 62, 4915-4921 (1994).

46. S.P. Hackett and D.L. Stevens. J Infect Dis, 168(1), 232-235 (1993).

47. V. Kappur, M.W. Majesky, L.L. Li, R.A. Black and J.M. Musser. Proc Natl Acad Sci USA, 90, 7676-7680 (1993).

48. D.L. Stevens. Pediatr Infect Dis J, 13, 561-566 (1994).

49. E.M. Adams, S. Gudmundsson, D.E. Yocum, R.C. Haselby, W.A. Craig and W.R. Sundstrom. Arch Intern Med, 145, 1020-1023 (1985).

50. P.R. Martin and E.A. Hoiby. Scand J Infect Dis, 22, 421-429 (1990).

51. W. Kohler. Zbl Bakt, 272, 257-264 (1990).

52. V. Hribalova. Ann Intern Med, 108, 772 (1988).

53. E.T. Gaworzewska and G. Coleman. N Engl J Med, 321(22), 1546 (1989).

54. H. Eagle. Am J Med, 13, 389-399 (1952).

55. D.L. Stevens, A.E. Gibbons, R. Bergstrom and V. Winn. J Infect Dis, 158, 23-28 (1988).

56. D.L. Stevens, A.E. Bryant and S. Yan. Int J Antimicrobial Agents, 4, 297-301 (1994).

57. D.L. Stevens, S. Yan and A.E. Bryant. J Infect Dis, 167, 1401-1405 (1993).

58. S. Yan, P.M. Mendelman and D.L. Stevens. FEMS Microbiol Letters, 110, 313-318 (1993).

59. S. Yan, G.A. Bohach and D.L. Stevens. J Infect Dis, 170, 609-614 (1994).

60. D.L. Stevens, K.A. Maier and J.E. Mitten. Antimicrob Agents Chemother, 31, 213-218 (1987).

61. C.G. Gemmell, P.K. Peterson, D. Schmeling, Y. Kim, J. Mathews, L. Wannamaker and P.G. Quie. J Clin Invest, 67, 1249-1256 (1981).

62. D.L. Stevens, A.E. Bryant and S.P. Hackett. Clin Infect Dis, 20, S154-S157 (1995).

63. A. Norby-Teglund, R. Kaul, D.E. Low, A. McGeer and M. Kotb. 35th Annual ICAAC, (1995).

64. W. Barry, L. Hudgins, S. Donta and E. Pesanti. JAMA, 267, 3315-3316 (1992).

65. J.M. Yong. Lancet, 343, 1427 (1994).

66. B. Stegmayr, S. Bjorck, S. Holm, J. Nisell, A. Rydvall and B. Settergren. Scand J Infect Dis, 24, 589-597 (1992).

67. R. Kaul, A. McGeer, A. Norby-Teglund, M. Kotb and D. Low. 35th Annual ICAAC, (1995).

68. T. Shimozato, M. Iwata and N. Tamura. Infect Immun, 58, 1384-1390 (1990).

69. A. Achiron, R. Margalit, R. Hershkoviz, D. Markovits, T. Reshef, E. Melamed, I.R. Cohen and O. Lider. J Clin Invest, 93, 600-605 (1994).

70. D.L. Stevens, A.E. Bryant, S.P. Hackett, A. Chang, G. Peer, S. Kosanke, T. Emerson and L. Hinshaw. J Infect Dis, 173, 619-626 (1996).

71. D.L. Stevens. Clin Infect Dis, 21, 977-980 (1995).

72. L.J. Greenberg, E.D. Gray and E. Yunis. J Exp Med, 141, 934-943 (1975).

73. L. Weinstein and M. Barza. N Engl J Med, 289, 1129 (1972).

74. J.B. Zabriskie, D. Lavenchy, R.C.J. Williams and et al. Arthritis Rherm, 28, 1047-1051 (1985).

DIAGNOSIS OF CHLAMYDIAL INFECTION IN THE PEDIATRIC POPULATION

M.R. Hammerschlag
Departments of Pediatrics and Medicine
State University of New York Health Science Center at Brooklyn
Brooklyn, New York 11203-2098

ABSTRACT

The genus *Chlamydia* now contains 4 species, 2 of which, *Chlamydia trachomatis* and *C. pneumoniae* are important human pathogens. Both organisms cause infections in children and adults, but infection in children pose a unique set of problems. As *C. trachomatis* is primarily a sexually transmitted disease, the presence of rectal or genital infection in a prepubertal child has been used as evidence of sexual abuse. Although there are several categories of non-culture tests tht have been approved for genital sites in adults, these tests are not approved for these sites in children. Of these tests in rectal and vaginal specimens in children have been associated with a high rate of false positives. *C. pneumoniae* is emerging as a frequent cause of community acquired pneumonia in adults and children. Because culture is not generally available, serologic diagnosis is used more frequently. However, currently available serologic methods appear to be insensitive in children. The availability of a commercial PCR test will greatly facilitate the diagnosis of *C. pneumoniae* infections in children.

INTRODUCTION

The genus *Chlamydia* is a group of obligate intracellular parasites with a unique developmental cycle with morphologically distinct infectious and reproductive forms. All members of the genus have a Gram-negative envelope without peptidoglycan, share a genus-specific lipopolysaccharide antigen, and utilize host adenosine triphosphate (ATP) for the synthesis for Chlamydial protein (1). The chlamydial developmental cycle involves an infectious, metabolically inactive extracellular form (elementary body) and a noninfectious, metabolically active intracellular form (reticulate body). Elementary bodies, which are 200 to 400 nm in diameter, attach to the host cell by a process of electrostatic binding and are taken into the cell by endocytosis that is not dependent on the microtubule system. Within the host cell, the elementary body remains within a membrane-lined phagosome. Fusion of the phagosome with the host cell lysosome does not occur. The elementary bodies then differentiate into reticulate bodies that undergo binary fission. After approximately 36 hours the reticulate bodies differentiate into elementary bodies. At about 48 hours, release may occur by cytoloysis or by a process of exocytosis or extrusion of the whole inclusion, leaving the host cell intact (1).

The genus now contains four species, *Chlamydia trachomatis*, *Chlamydia psittaci*, *Chlamydia pneumoniae* (TWAR strain) and *Chlamydia pecorum*. The first three species are capable of causing infection in humans. *C. pecorum* was recently speciated off from *C. psittaci* and infects cattle and sheep, human infection has not as yet been described (2). The routes of transmission, susceptible populations, and clinical presentations differ markedly for the three species that cause infection in humans.

Most chlamydial infections in children are due to *C. trachomatis* or *C. pneumoniae*. *C. trachomatis* infection is acquired perinatally by infants during partuition (3). Approximately 50 to 70% of infants born to chlamydia-infected women will become infected at one or more anatomic sites, including the conjunctiva, nasopharynx rectum and vagina. Clinically, infants will develop conjunctivitis and/or pneumonia. The nasopharynx is the most frequently infected site, most infections are asymptomatic and may persist for at least 3 years. Rectal and vaginal infections are usually asymptomatic and may also persist for at least 3 years. Rectal and vaginal infection in older children may also be acquired through sexual abuse.

C. pneumoniae is emerging as a common respiratory pathogen in children as well as adults. Transmission is thought to be person-to-person through infected respiratory droplets. Preliminary studies suggest that *C. pneumoniae* may be responsible for 8 to 15% of community-acquired pneumonia in children over 6 mos of age (4).

DIAGNOSIS OF *C. TRACHOMATIS* INFECTIONS IN CHILDREN

The "gold standard" is isolation by culture of *C. trachomatis* from the conjunctiva, nasopharynx, vagina or rectum. Chlamydia culture has been further defined by the Centers for Disease Control as isolation of the organism in tissue culture and confirmation by microscopic identification of the characteristic inclusions by fluorescent antibody staining (5). Several nonculture methods have FDA approval for diagnosis of chlamydial conjunctivitis. They include enzyme immunoassays (EIA), specifically Chlamydiazyme (Abbott Diagnosistics, North Chicago, IL), Pathfinder (Sanofi-Pasteur, Chaska, MN) and SureCell (Kodak, Rochester, NY) and direct fluorescent antibody tests (DFA) including Syva MIcroTrak (Genetic Systems, Seattle, WA) and Pathfinder (Sanofi-Pasteur, Chaska, MN). These tests appear to perform very well with conjunctival specimens with sensitivities $\geq$ 90% and specificities $\geq$ 95% compared to culture (3). Unfortunately, the performance with nasopharyngeal specimens has not been as good with sensitivities ranging from >90% in infants with pneumonia to 33% to 87% in nasopharyngeal specimens in infants with conjunctivitis. The commercially available DNA probe, Pace II (GenProbe, San Diego, CA) has FDA approval only for cervical and urethral sites in adults, where its performance has been similar to most of the approved EIAs available. It does not have approval for any site in children, including the conjunctiva, vagina or rectum. The recently approved polymerase chain reaction (PCR) assay, Amplicor (Roche Molecular Systems, Branchburg, NJ), also has approval only for genital sites in adults. However, preliminary data suggest that PCR is equivalent to culture for detection of *C. trachomatis* in the conjunctiva and possibly superior to culture for detection of the organism in nasopharyngeal specimens (6).

Nonculture tests should never be used for rectal or vaginal sites in children, or for any forensic purposes in adolscents an adults. This is stated clearly in the 1993 Chlamydia Guidelines and the 1993 STD Treatment

Guidelines (5,7). Use of these tests for vaginal and rectal specimens has been associated with a large number of false positive test (8,9,10). Fecal material can give false-positive reactions with any EIA; none are approved for this site in adults. Common bowel organisms, including *E. coli*, Proteus ssp., vaginal organisms such as Group B Streptococcus and *Gardnerella vaginalis* and even some respiratory flora such as Group A Streptococcus can also give positive reactions with EIAs (10). These types of test are best for screening for genital infection in adolescents and adults in high prevalence populations (prevalence of infection >7%). There are very few reports on the performance of the DNA probe, but it appears to be equivalent to most available EIAs, in terms of sensitivity and specificity compared to culture for genital specimens. Another potential problem can occur with use of an EIA for respiratory specimens. As all of the available EIAs use genus-specific antibodies, if used for respiratory specimens, these tests will also detect *C. pneumoniae* (11).

Even though culture is considered the gold standard, culture of *C. trachomatis* is not regulated in any way, sensitivity may vary from laboratory to laboratory (12). The methods used for culture confirmation became an issue when several large commercial laboratories started using an EIA instead of FA staining and visual identification of inclusions for culture confirmation. This has resulted in at least one "outbreak" of *C. trachomatis* infection in the evaluation of suspected sexual abuse among residents and staff of an institution for the mentally retarded in Ohio in 1990 (13). All the "positive" cultures, mostly rectal specimens, were subsequently determined to be false-positives resulting from carry over of fecal material and bacteria in the culture specimens. The major advantage of culture is that it is 100% specific. When cultures are obtained for *C. trachomatis* in the evaluation of suspected sexual abuse, one should pay careful attention to the laboratory used. Unlike Canada, we do not have a system of designated reference laboratories.

DIAGNOSIS OF *C. PNEUMONIAE* INFECTIONS IN CHILDREN

A specific laboratory diagnosis of *C. pneumoniae* infection can be made by isolation of the organism from nasopharyngeal or throat swabs, sputa or pleural fluid, if present. The nasopharynx appears to be the optimum site for isolation of the organism (4). The relative yield from throat swabs and sputum is not known.

Initial studies suggested that *C. pneumoniae* was very difficult to isolate in tissue culture compared to *C. trachomatis*. Originally the same methods were used, HeLa or McCoy cells pretreated with dextran diethyl-aminoethyl (DEAE). Multiple passages were needed, the inclusions were very small and difficult to see and, in general, the yield was very poor. *C. pneumoniae* grows more readily in other cell lines derived from respiratory tisue, specifically HEp-2 and HL cells (14). Omission of pretreatment with dextran DEAE results in much larger inclusions and specimens need only be passed once. Culture with an initial inoculation and one passage should take 3 to 7 days.

Nasopharyngeal cultures can be obtained with Dacron-tipped wire shafted swabs. Specimens for culture should be placed in appropriate transport media, usually a sucrose-phosphate buffer with antibiotics and fetal calf serum, and stored immediately at 4 C for no longer than 24 hours. Viability decreases if specimens are held at room temperature. If the specimen cannot be processed within 24 hours they should be frozen at -70 C until culture can be performed. After 72 hours incubation, culture confirmation can be performed by staining with either a *C. pneumoniae* species-specific or a *Chlamydia* genus-specific (anti-LPS) fluorescein-conjugated monoclonal antibody (15). Inclusions of *C. pneumoniae* do not contain glycogen, thus will not stain with iodine. Unfortunately, there is limited availability of commercially produced *C. pneumoniae*-specific reagents. If a genus-specific antibody is used, *C. pneumoniae* should be confirmed by differential staining with a specific *C. trachomatis* antibody; if the latter is negative, then the isolate is either *C. pneumoniae* or *C. psittaci*. If there was no avian exposure, psittacosis would be highly unlikely.

As isolation of *C. pneumoniae* was difficult and initially limited, more emphasis was placed on serologic diagnosis. However, performance of the microimmunofluorescence (MIF) test is also limited to a small number of research laboratories. The MIF was modified from the test used for *C. trachomatis* by using EBs from TW 183 or other *C. pneumonie* strains as the antigen. With the MIF test one can detect IgG, IgM and IgA antibodies. Grayston and colleagues have proposed a set of criteria for serologic diagnosis of *C. pneumoniae* infection with the MIF test that is used by many laboratories and clinicians. For acute infection the patient should have a four-fold rise in IgG titer, a single IgM titer of 1:16 or greater, or a single IgG titer of 1:512 or higher. Past or pre-existing infection is defined as an IgG titer of 1:16 or higher but less than 1:512. It was further proposed that the pattern of antibody response in primary infection differed from that seen in reinfection. In initial infection, the IgM response appears about 3 weeks after the onset of illness and the IgM response appears at 6 to 8 weeks. In reinfection, the IgM response may be absent and the IgG occurs earlier, within 1 t o 2 weeks. A four-fold titer rise or a titer ≥1:64 with the CF test is also felt to be diagnostic. Initially, Grayston and colleagues found that fewer than one third of hospitalized patients with suspected *C. pneumoniae* infection had detectable CF antibody. However, in a recent report of a small outbreak of *C. pneumoniae* infections among University of Washington students, all seven patients with pneumonia had CF titers of 1:64 or greater (18). The CF test is genus specific.

Because of the relatively long period until the development of a serologic response in primary infection, the antibody response may be missed if convalescent sera are obtained too soon, i.e., earlier than 3 weeks after the onset of illness. Use of paired sera also only affords a retrospective diagnosis, which is of little help in terms of deciding how to treat the patient. The criteria for use of a single serum sample have not been correlated with the results of culture and are based mainly on data from adults. The antibody response in acute infection may take longer than 3 months to develop. Acute, culture-documented infection can also occur without seroconversion, especially in children (4,19,20). Only 28% of the culture-positive children enrolled in the multicenter pneumonia treatment study met the serologic criteria for acute infection; most had no detectable antibody by the MIF test even after 3 months of followup (4). However, the results of immunoblotting revealed that these children have antibody to a number of *C. pneumoniae* proteins but that fewer than 30% react with the major outer membrane protein (MOMP), which is the antigen presented in the MIF test (21). Although the MOMP has been demonstrated to be immunodominant in *C trachomatis* infection, it does not appear to be for *C. pneumoniae* (22,23).

Direct detection of *C. pneumoniae* EBs in clinical specimens by fluorescent antibody stains is sometimes possible, but is insensitive and frequently nonspecific (17). There are also no commercially available reagents that have been evaluated or approved for this purpose. All the currently available chlamydial EIAs will detect *C. pneumoniae* as well as *C. trachomatis* since they use polyclonal or genus-specific monoclonal antibodies. However, there are few data on the use of these assays in this setting and data that are available also suggest that EIAs are insensitive for detection of *C. pneumoniae* in respiratory specimens. Chirgwin et al (19) obtained nasopharyngeal from 91 patients with pneumonia for testing with Chlamydiazyme. Although there were no false positives, the EIA detected only 2 of 15 (15%) patients who were culture positive for *C. pneumoniae*.

The number of *C. pneumoniae* organisms present in the respiratory tract of individuals with pneumonia or other respiratory disease are fewer than the number found in genital *C. trachomatis* infection. DNA amplification methods, e.g., PCR, appears to be the most promising technology in the development of a rapid, nonculture method for detection of *C. pneumoniae*. Although there are no commercially available kits, several investigators have evaluated in-house PCRs. In general, PCR appears to be at least as sensitive as culture for detecting *C. pneumoniae* in throat and nasopharyngeal specimens (24-27).

REFERENCES

1. J.W. Moulder. Am Soc Microbiol 55;143-190 (1991).
2. H. Fukushi and K. Hirai. Microbiol Immunol 37516-522 (1993).
3. M.R. Hammerschlag. Pediatr Ann 23,349-53 (1994).
4. S. Block, J. Hedrick, M.R. Hammerschlag, G.H. Cassell and J.C. Craft. Pediatr Infect Dis J 14,471-7 (1995).
5. Centers for Disease Control. MMWR 42, No. RR-12 (1993).
6. P.M. Roblin, N. Sokolovskaya, M. Gelling, N. Tsumura, J. Jules, A. Kutlin and M.R. Hammerschlag. Pediatr Res 39,301A (1996).
7. Centers for Disease Control. MMWR 42, No. RR-14 (1993).
8. M.R. Hammerschlag, P.J. Rettig, M.E. Shields. Pediatr Infect Dis J 7,11-14 (1988).
9. K. Porder, N. Sanchez, P.M. Roblin, M. McHugh and M.R. Hammerschlag. Pediatr Infect Dis J 8,358-60 (1989).
10. S.B. Hauger, J. Brown, F. Agre, F. Sahraie, R. Ortiz and P. Ellner. Pediatr Infect Dis J 7,11-14 (1988).
11. J.E. Bauwens, M.S. Gibbons, M.M. Hubbard and W.E. Stamm. J Pediatr 119,591-3 (1991).
12. M.S. Pate and E.W. Hook, 3rd. Sex Transm Dis 22,322-6 (1995).
13. Centers for Disease Control. MMWR 39,932-5 (1991).
14. P.M. Roblin, W. Dumornay and M.R. Hammerschlag. J Clin Microbiol 30,1968-71 (1992).
15. G.S. Montalban, P.M. Roblin and M.R. Hammerschlag. J Clin Microbiol 32,1406-7 (1994).

16. J.T. Grayston, L.A. Campbell, C-C Kuo, C.H. Mordhorst, P. Saikku, D.H. Thom and S-P Wang. J Infect Dis 161,618-25 (1990).
17. J.T. Grayston, C-C Kuo, S-P Wang and T. Altman. N Engl J Med 315,161-8 (1986).
18. J.T. Grayston, M.B. Aldous, A. Easton, S-P Wang, C-C Kkuo, L.A. Campbell and J. Altman. J Infect Dis 168,1231-5 (1993).
19. K. Chirgwin, P.M. Roblin, M. Gelling, M.R. Hammerschlag and J. Schachter. J Infect Dis 163,757-61 (1991).
20. U. Emre, P.M. Roblin, M. Gelling, W. Dumornay, M. Rao, M.R. Hammerschlag and J. Schachter. Arch Pediatr Adolesc Med 148,727-32 (1994).
21. A. Kutlin, P.M. Roblin, M.R. Hammerschlag. Unpublished data.
22. C.M. Black, J.E. Johnson, E.E. Farshy, T.M. Brown and B.B. Berdol. J Clin Microbiol 29,1312-6 (1991).
23. Y. Iijima, N. Miyashita, and T. Kishimoto. J Clin Microbiol 32,583-8 (1994).
24. L.A. Campbell, M.P. Melgosa, D.J. Hamilton, C-C Kkuo and J.T. Grayston. J Clin Microbiol 30,434-9 (1992).
25. C.A. Gaydos, C.L. Lowler, V.J. Gill, J.T. Eiden and T.C Quinn. Clin Infect Dis 17,718-23 (1993).
26. C.A. Gaydos, P.M. Roblin, M.R. Hammerschlag, C.L. Hyman, J.J. Eiden, J. Schachter, T.C. Quinn. J Clin Microbiol 32,903-5 (1994).
27. C.A. Gaydos, J.J. Eiden, D. Oldach, L.M. Mundy, P. Auwalter, M.L. Warner, E. Vance, A.A. Burton and T.C. Quinn. Clin Infect Dis 19,157-60 (1994).

DFA, EIA, PCR, LCR AND OTHER TECHNOLOGIES: WHAT TESTS SHOULD BE USED FOR DIAGNOSIS OF CHLAMYDIA INFECTIONS?

J. Schachter
Department of Laboratory Medicine
University of California, San Francisco
Box 0842
San Francisco, CA 94143-0842

ABSTRACT

For many years, isolation in tissue culture (TC) was considered the test of choice for diagnosis of *Chlamydia trachomatis* infection. Non-culture tests, such as direct fluorescent antibody (DFA) and enzyme immunoassay (EIA) which detected chlamydial antigens in clinical specimens, made chlamydia diagnostic tests more widely available. DFA and EIA were less sensitive than TC and had some false positive results which compromised our ability to use these tests in low prevalence settings.

Direct nucleic acid probes are available, but do not appear to be more sensitive than EIA. It was only with the introduction of amplified DNA tests [polymerase chain reaction (PCR) and ligase chain reaction (LCR)] that non-culture tests became available that were actually more sensitive than TC. Unfortunately these tests are also more expensive than the antigen detection methods. Until there is a fairly sophisticated cost benefit analysis or a change in the pricing of these tests, it seems obvious that TC will remain the best choice where medical/legal implications are important, DFA will probably remain a widely used tests for laboratories that process relatively small numbers of specimens and EIAs will play a role where cost is major factor and large numbers of specimens require bulk processing. Where they are affordable, the amplified DNA tests are to be preferred as they are far more sensitive than these other non-culture tests.

INTRODUCTION

Choice of which diagnostic test for *Chlamydia trachomatis* infection will be used by a specific laboratory depends on more than the performance profile of the particular test in question. Cost, of course, is an issue. The population which is being tested will factor into the final decision, and the volume of tests to be processed by the laboratory will also be a factor.

The oldest diagnostic test for chlamydial infection is the Giemsa stain. This method of demonstrating inclusions in infected epithelial cells was used for 50 years before the trachoma organism was isolated. This is a procedure which is not recommended for routine use on genital tract specimens. The problems are sensitivity and adequate specimen collection, as the specimens must contain a large number (thousands) of epithelial cells. Evaluation of a single slide is quite time consuming, and this procedure has been shown to be of poor sensitivity in detecting chlamydial infection of the male urethra, being positive in only 15% of culture proven infections. The only place where the Giemsa stain can be regularly recommended is in acute and severe inclusion conjunctivitis of the newborn, where the infections are accompanied by large amounts of agent and inclusions are usually easily found.

Fluorescein-conjugated monoclonal antibodies directed against a species-specific epitope of *Chlamydia trachomatis* are commercially available.(1) The antibodies directed against the species-specific epitope on the major outer membrane protein (MOMP) are preferred over the genus-specific LPS target. The quality of fluorescence appears more homogenous and intense with the antibodies against MOMP, which is a dominant protein in the outer membrane of the organism, evenly distributed across it's surface. The DFA (direct fluorescent antibody) test stains elementary bodies in epithelial cell scrapings from infected sites. DFA is less sensitive than culture in detecting chlamydial infection but is a relatively specific test. The procedure seems best suited for use in high risk populations, and where culture is not available.

Enzyme immunoassay (EIA) for detection of chlamydial antigen in clinical specimens is also available.(2) This test appears to have a similar sensitivity/specificity profile as the fluorescent antibody test. The test will probably be best aimed at the same populations (high prevalence settings where culture is not available).

These two non-culture methods each have advantages and disadvantages. They require special equipment (spectrophotometer or FA microscope). Experienced microscopists are required for the FA procedure, EIA is more suitable for batching and FA is more convenient for testing small numbers of specimens and is faster. Neither requires rigorous maintenance of a cold chain for specimen shipment and both provide results faster than culture.

Since their introduction these tests have been improved. The sensitivity for cervical infection is about 75 - 80% (lower for the male urethra) and specificity about 98% for detecting chlamydiae. Blocking tests to confirm positive reactions can improve EIA specificity to >99%, thus making screening of lower prevalence populations more feasible.(3)

The direct gene probe is a more recent product test. It has not been as well evaluated, but the performance seems to be similar to the EIAs.(4) The 1 - 2% false positive rate seen with DFA procedures precludes their use in screening low-prevalence populations. EIAs with confirmatory tests, which result in specificities of 99.7% or greater will provide a high enough predictive value for results to be accepted in the low-prevalence setting.

Amplified DNA probe technology, exemplified by PCR and LCR, is the newest technology coming to market. PCR appears to be very sensitive in diagnosing chlamydial infection of the male urethra and in detecting chlamydial genes in male urine specimens, but appears to be less sensitive in diagnosis of cervical infection due to inhibitors in the specimens.[5,6] LCR technology is now commercially available. It appears to be less susceptible to inhibitors and is thus slightly more sensitive than PCR.[7-9] Both of these tests can be applied for detection of chlamydial nucleic acid in first catch urine specimens from men and from women. If this is borne out with subsequent experience, this may well revolutionize chlamydial diagnostics by allowing for non-invasive screening programs to form the basis of chlamydia control programs.

PCR and LCR are the first commercially available tests that are more sensitive than isolation in cell culture in the hands of expert laboratories. All previous tests were less sensitive than isolation. It should be noted that the sensitivity of tissue culture systems vary dramatically but it would seem, predictably, that the technically less demanding PCR and LCR tests perform in a similar manner in different types of laboratory settings.

All of these tests take time to perform and the results are seldom available for at least another day. The current generation of rapid (patient side, or point of care) diagnostic tests is not recommended because of shortcomings in either sensitivity or specificity.

SEROLOGY

It should be noted that there are no serologic tests recommended for routine diagnosis of genital tract infection with *C. trachomatis*, although sensitive and specific tests to measure antichlamydial antibodies are available.[10]

ISOLATION PROCEDURES

Isolation of the agent from a patient's tissue has long been considered the definitive test.[11] However, it is not 100% sensitive, but only amplified DNA assays are more sensitive. Culture misses can result from sampling variation and poor specimen handling.

A number of tissue culture isolation procedures have been developed. The initial procedure involved irradiated McCoy cells. The irradiation procedure has been supplanted by treatment of the cell monolayer with antimetabolites which provide the chlamydia some competitive advantage and allow higher infection rates. The basic step, crucial to all the isolation procedures, is the centrifugation of the inoculum into the tissue culture monolayer. The cells are grown on coverslips which are inoculated, incubated at 35°C for 48-72 hours, and then stained by either Giemsa, iodine or fluorescent antibody procedures, and then examined microscopically for inclusions.

The method of choice uses cycloheximide-treated McCoy cells. There is nothing unique about the McCoy cells, since BHK-21 and other cells have been shown to be susceptible to chlamydial infections. Some workers with long experience in this particular system have used DEAE-treated HeLa cells quite successfully.

The most sensitive isolation procedure involves inoculation of clinical specimens into 1 dram vials, a blind passage, and the use of fluorescent antibodies to stain the inclusions.

THE GOLD STANDARD CHANGES - AND SO DO THE TESTS

It should be noted that the gold standard for evaluating chlamydia diagnostic tests has changed over the past 10 - 15 years. Initially, isolation in cell culture was considered to be the gold standard, but it was soon recognized that some specimens that were EIA and DFA positive were culture negative. Culture in most expert laboratories is about 75 - 85% sensitive. EIA and DFA procedures are less sensitive than culture with the typical result being on the order of 75 - 80% for cervical specimens and 60 - 70% for specimens from the male urethra. These tests, in general, are much less sensitive than culture although some of the more recently introduced EIAs approach sensitivity of TC.

Cell culture is still a specialized procedure that is typically performed in the context of research studies and is obviously more sensitive that EIA or DFA. In fact, until the introduction of amplified DNA technologies, TC was clearly the most sensitive procedure available. Thus, it was relatively easy to make recommendations as to which tests to use. For optimal sensitivity in the past culture was the test of choice - DFA was best suited for low volume laboratories and moderate to high-prevalence settings while the EIAs were reasonable choices for use in moderate to high-prevalence populations. With use of confirmatory tests EIA could be extended to low-prevalence populations with high volume laboratories where batch processing was possible. The number of control specimens required for EIAs made them impractical and too expensive for use in a low volume laboratory.

The introduction of direct DNA probes really did not change the approaches to diagnosis. There is only one commercially available test that is quite specific, but seems no more sensitive than the EIAs.

The introduction of amplified DNA technology has really placed laboratory workers in an awkward position. These tests are far more expensive than EIA and DFA, but they are also far more sensitive. Their specificity is very high and there is no need for confirmatory tests as the specificity is on the order of 99.8% or greater and thus the predictive value of a positive result will be very high in all settings where screening could be practically considered. It is clear that these tests are more sensitive than culture and indeed in comparative evaluations (for those specimens where inhibitors are not a problem) the LCR and PCR tests have generated 25 - 35% more positive results than were obtained by culture. Predictably, that increment is even higher when these tests are compared to antigen detection methods.

Thus, there is a trade-off. It costs more money to use PCR and LCR, but the yield is greater. These tests are, however, not perfect. Even in the best of settings approximately 5% of specimens are culture positive and negative by amplified DNA tests. This is likely due to the presence of inhibitors in at least some of those falsely negative specimens. Thus, if one is in a situation where it is important to identify as many infected individuals as possible, the laboratory must use an amplified DNA test plus culture. The amplified DNA tests are, by far, the tests of choice from all technical considerations but it is likely the EIAs will still continue to be used in some settings because they are far less expensive.

Another factor in the decision of which test to use will be which specimens are to be processed. One major advantage promised by the amplified DNA test is their ability to detect chlamydial genes in urine specimens. This test is more sensitive than culture for cervical specimens and LCR on urine will generate at least as many positive results as LCR on cervical swabs will in women, and urine PCR and LCR will yield more than urethral swabs in men. Thus, it is likely in screening asymptomatic individuals where urethral swabs are unlikely to be collected from many men and where it may not be necessary to perform pelvic examinations on all women, that the amplified DNA test applied to a first-catch urine specimen will be the only option. This ability to detect chlamydial genes in urine specimens is a major diagnostic advance and promises to be the basis of screening programs used in chlamydia control programs.

REFERENCES

1. Tam MR, Stamm WE, Handsfield HH, Stephens R, Kuo C-C, Holmes KK, Ditzenberger K, Crieger M, Nowinski RC. N Engl J Med 310(18):1146-1150, 1984.

2. Schachter J. Yale J Biol Med 58:443-452, 1985.

3. Moncada J, Schachter J, Bolan G, Engelman J, Howard L, Mushahwar I, Ridgway G, Mumtaz G, Stamm W, Clark A. J Clin Microbiol 28(8):1770-1773, 1990.

4. Clarke LM, Sierra MF, Daidone BJ, Lopez N, Covino JM, McCormack WM. J Clin Microbiol 31(4):968-971, 1993.

5. Jaschek G, Gaydos CA, Welsh LE, Quinn TC. J Clin Microbiol 31(5):1209-12, 1993.

6. Bauwens JE, Clark AM, Stamm WE. J Clin Microbiol 31(11):3023-3027, 1993.

7. Schachter J, Stamm WE, Quinn TC, Andrews WW, Burczak JD, Lee HH. J Clin Microbiol 32(10):2540-2543, 1994.

8. Chernesky MA, Lee H, Schachter J, Burczak JD, Stamm WE, McCormack WM, Quinn TC. J Infect Dis 170(5):1308-1311, 1994.

9. Lee HH, Chernesky MA, Schachter J, Burczak JD, Andrews WW, Muldoon S, Leckie G, Stamm WE. Lancet 345(8944):213-6, 1995.

10. Wang S-P, Grayston JT. J Infect Dis 130:388-397, 1974.

11. Stamm WE. Ann Intern Med 108(5):710, 1988.

MOLECULAR DIAGNOSIS OF HELICOBACTER PYLORI

T. Ulf Westblom
Department of Internal Medicine
Texas A&M University, College of Medicine
Department of Veterans Affairs
Central Texas Veterans Health Care System
Temple, TX 76504

ABSTRACT

H. pylori infection can be diagnosed with many different tests. If the patient is undergoing endoscopy with gastric biopsies, culture and histology remain the diagnostic methods of choice. Indirect tests include rapid urease tests, urea breath tests, and serology. Molecular methods such as PCR offer marginal improvements when done on biopsy material, but has the advantage of being able to accurately identify H. pylori in areas outside the stomach where cultures usually fail. PCR can detect low numbers of organisms in gastric juice, bile, stool and oral secretions. Because of its high sensitivity it can also be used for epidemiologic investigations of environmental sources. However, the largest role for PCR may be in molecular fingerprinting. Arbitrary Primer PCR (RAPD) on the whole bacterial genome can reliably and accurately distinguish between isolates. PCR-based RFLP analysis can separate isolates based on restriction fragment sizes in a smaller amplified genome segment. REP-PCR can group isolates into clusters that appear to have different clinical expressions. These methods promise to shed new light on the transmission and pathogenicity of H. pylori.

BACKGROUND

Success in diagnosing a disease depends to a large extent upon the choice of diagnostic techniques. Nowhere is this more obvious than in the case of gastroduodenal infection with Helicobacter pylori. This spiral-shaped bacterium had been observed in the gastric mucosa by several independent investigators going back to the beginning of the century (1-4). Yet, their existence was often doubted by the experts of the time. In 1954 Palmer reported a large study of more than 1000 patients undergoing suction biopsies of their stomachs (5).

Palmer had specifically looked for the "spirochetes" observed by others, but found none and concluded that the organisms might be "simple contamination of the mucosal surface by swallowed spirochetes ... encouraged by certain gastric diseases and is common at death, but actual invasion of the tubule lumina and of the tissue does not occur during life."

Palmer missed his big opportunity to discover H. pylori by using an inappropriate diagnostic test. He stained his biopsies with the commonly used H&E stain. This stain is excellent for displaying tissue morphology, but is a very poor stain for visualizing H. pylori. Twenty-five years later Robin Warren in Australia was looking at gastric mucosa using a Warthin-Starry silver stain (6). This stain shows the bacteria very well and soon he and Barry Marshall would surprise the scientific community by reporting the presence of "Unidentified curved bacilli on gastric epithelium in active chronic gastritis" (7). In the years that followed several diagnostic tests specific for H. pylori were developed. Recently advances in molecular techniques have been applied to this organism resulting in very specific and sophisticated diagnostic tools.

CONVENTIONAL METHODS OF DIAGNOSIS

In the early days after the discovery of H. pylori there were only two ways of diagnosing H. pylori. You could use a histologic stain, and if you had a good laboratory you might also be able to culture the organism. Warren used the Warthin-Starry stain in his first studies, but soon the cheaper and quicker Giemsa stain became the stain of choice (8). If simultaneous visualization of the tissue morphology is desired the Genta stain is often used today (9).

H. pylori is a fastidious organism and does not grow well on conventional culture media. In a humid CO_2-containing atmosphere it can be retrieved on rich agars containing blood or egg yolk or in selective liquid media containing serum (10). Culture diagnosis is valuable since it can help with further characterization of the isolate such as antibiotic resistance or presence of pathologic markers. A draw-back is its relatively low sensitivity, only 80-85% in most laboratories compared to around 95% for histology.

Several indirect diagnostic tests have been developed based on the powerful urease enzyme of H. pylori. Urease converts urea to CO_2 and ammonia, a reaction that helps H. pylori neutralize the gastric acid in its immediate surrounding. The same reaction can be spotted with the help of an indicator like phenol red and be used as a rapid test for H. pylori (11). If the patient is allowed to swallow a cocktail containing nuclear tagged urea the same reaction can be identified by measuring the amount of radioactive CO_2 exhaled. The Urea Breath Test has a high specificity and requires no endoscopy for gastric biopsy (12,13). A minor draw-back is its tendency for false negative tests if the patient has consumed any drugs in the preceding 2 weeks that could have suppressed the growth of the bacteria (14,15).

Serology has emerged as a popular noninvasive test for H. pylori. There are a multitude of commercial tests available on the market. The ELISA-based tests usually have good sensitivity and specificity around 95% (16) while rapid antibody kits have proved less reliable (17,18). An ELISA test can be an acceptable way of making an initial diagnosis of H. pylori infection, but can not reliably be used to follow the outcome of therapeutic interventions. Since titers can take six months or more to fall significantly other tests have to be used after treatment.

PCR FOR CLINICAL DIAGNOSIS

Molecular tests offer excellent ways for very precise diagnosis of H. pylori infection. The techniques involved may not be readily available at every laboratory, but the tests do not require the bacteria to be alive when tested. This means that samples can be shipped between institutions without compromising the results of the tests (19). Polymerase Chain Reaction (PCR) is an excellent method for diagnosing H. pylori when the organism is present in low numbers. Theoretically the method can find and identify an organism if only a few single copies of its DNA is present. The DNA molecule is very stable chemically and can survive in the environment for long periods of time (20). This makes the method suitable for both clinical and environmental sampling. The extreme sensitivity of the test can be a problem though. Any laboratory that wishes to perform

PCR for diagnostic purposes must develop strict guidelines for the handling of clinical specimens (21,22). Areas that receive specimens and prepare them for PCR should be physically separated from areas handling the DNA amplification and the visualization of amplified DNA products.

Several PCR protocols for clinical diagnosis of H. pylori have been published. They differ from each other mostly in the choice of primers. The first PCR protocol to be published used primers from the 16S rRNA (23). This is an approach that is well established from work with other bacteria and many other investigators have followed their lead (24-28). Since 16S ribosomal RNA is extensively studied for a large number of organisms it is easy to avoid areas of likely cross-reaction. The potential disadvantage is that genes can be relatively conserved in many areas since 16S ribosomal RNA is found in most living cells and phylogenetically has common roots. An area that appears to be unique may still have significant sequence homology with genes in bacteria that have not yet been fully characterized.

Other investigators have instead decided to use primers from the genes encoding the uniquely powerful urease of H. pylori (19,29-35). This approach has the potential for being more risky since many bacteria have urease genes and conserved regions among species can be expected. Still, it has been shown that careful selection of the primers can result in 100% specificity when tested against other urease-containing bacteria, including other Helicobacter species (19). An alternative approach was chosen by Hammar et al. (36) who used sequences from a protein antigen that appears to be species specific for H. pylori (37). Finally, Valentine et al. picked primers from a cloned fragment randomly selected from an H. pylori cloned library (38). Using a probe from the fragment they showed that it had high specificity for H. pylori. Selecting primers to amplify a 203-base-pair product from this fragment, they were able to achieve 100% specificity when tested on related bacteria.

All of these PCR protocols have proven themselves to be accurate in diagnosing H. pylori from clinical biopsy material. However, this may not be the optimal usage for this technique. When used on biopsy material PCR usually does not add much to other techniques (33,34,39). Culture and histology, when used together, detects H. pylori in almost as many cases as PCR. The exception would be in patients where the organism may be present only in very low numbers following attempts of eradication (40,41). The main advantage of PCR is instead in situations where you need to make a clinical diagnosis but don't have access to biopsy material. Westblom et al. studied the use of PCR on gastric juice which can be collected through a nasogastric catheter (19). Using only 5 ml of aspirated gastric juice they could correctly diagnose H. pylori infection with a specificity of 100% and a sensitivity of 96%. These findings have later been verified in at least two studies showing similar degrees of sensitivity and specificity (26,35). This compares very favorably to previous attempts to culture H. pylori from gastric juice which had uniformly poor results (42-45). H. pylori has also been detected in a similar way in bile samples using PCR (32). In the future it may be possible to use PCR to diagnose H. pylori from oral secretions or stool samples, bypassing the need for endoscopy. But those assays are far from standardized and we don't know enough yet about the relation between carrier states and true infections for them to be useful at this point.

PCR FOR MOLECULAR EPIDEMIOLOGY

PCR can be used to look for the presence of H. pylori outside the human stomach, including the environment. This makes it a useful tool for epidemiologic studies. An unanswered question is whether H. pylori is transmitted through oral-oral or fecal-oral route. Cultures from the oral cavity have almost uniformly been negative for H. pylori (46-48), but PCR has been able to find the organism in the oral cavity (26,49-51). Still, in some studies it has been present in only low numbers (26,36,52) challenging the significance of the finding. PCR has also been used to diagnose the presence of H. pylori in stool samples (53-55). This is another area where PCR offers distinct advantages since H. pylori may only be present in low numbers and the multitude of contaminating organisms in the bowel makes it very hard to obtain any positive cultures (46,56-58).

Two epidemiologic studies in South America have suggested that infected water supply may be a vehicle for transmission of H. pylori. Among children living in Lima, Peru, H. pylori infection was three times as common if the homes had an external water source than if it was internal. This was independent of the socioeconomic status of the families since both types of water sources could be found within the same neighborhoods (59). In Chile risk factors for H. pylori infection include consumption of uncooked vegetables and uncooked shell-fish (60). This suggests that contaminated water used for irrigation purposes may be an important factor in these countries.

To test the hypothesis that fecally contaminated water contains H. pylori Westblom et al. investigated samples of sewage water from Peru (61). Twenty 40 ml samples of pooled raw or partially filtered sewage water were ultracentrifugated and DNA was extracted from the pellets. Using previously validated primers from the UreA gene (19) they showed that 15% of the water samples contained H. pylori.

MOLECULAR FINGERPRINTING BY ARBITRARY PRIMER PCR

One area where PCR has been used with particular success is in molecular fingerprinting. This technique can differentiate between different strains and is helpful in determining if continued infection is due to treatment failure or reinfection (62,63). It can also help identifying transmission of H. pylori between patients or within families (64,65).

H. pylori is unique among bacterial pathogens in its extreme genetic diversity. This was first shown by Majewski and Goodwin using restriction endonuclease digestion (66). They found unique patterns among isolates from 70 patients. When pairs of isolates from 11 patients were compared the second isolate was markedly different in six patients. H. pylori fingerprinting by PCR was first done by Akopyanz et al. using Arbitrary Primer PCR (67). This technique uses single primers that have been arbitrarily chosen without regard to any specific genomic sequence. By using very low stringency parameters in the early PCR cycles these arbitrary primers amplify fragments from genomic sites to which they are fortuitously matched. After a subsequent series of cycles with high stringency a unique pattern of fragments are produced. This technique also found that H. pylori has an extreme genomic diversity. Of 64 isolates from different patients each one was distinguishable from each other. However, in three patients where multiple isolates had been obtained over a 6 month period all isolates from the same patient were identical.

Arbitrary primer PCR is now widely used as a sensitive and reliable method for fingerprinting of H. pylori (63,65,68,69). However, it has its limitations. You need the whole bacterial genome as a starting point which means you need to isolate the organism in pure culture first. This may not always be possible, particularly if you are dealing with archival material trying to make an identification after the fact. In those situations it is helpful to instead use the technique of RFLP analysis of PCR produced products. This was first described by Akopyanz et al. who amplified segments from the UreA-UreB genes and the neighboring UreC-UreD genes (70). These PCR-produced products were then digested with Mbol and HaeIII and observed for restriction fragment length polymorphism (RFLP). Using both of the selected gene segments 59 of 60 isolates could be separated from each other while repeat isolates from the same patient remained identical.

Using the same techniques, sometimes looking at other amplified gene segments, other investigators have verified that it is a very useful fingerprinting tool (71-75). The main advantage is that it does not require successful isolation of H. pylori prior to testing. A drawback is that it can be less discriminatory than arbitrary primer PCR, specifically if a gene with low heterogeneity is used (76). Another problem is if more than one strain of H. pylori is present in the material. The restriction fragments from the different strains would then mix together producing a pattern that corresponds to none of the strains. However this can be detected by carefully adding together the lengths of each fragment. If the sum of them markedly exceeds the size of the whole gene fragment more than one strain must have been present. PCR-based RFLP can therefore be used to prove the existance of mixed populations of H. pylori (74).

Typing of isolates based on fingerprints may also help to shed light on whether some strains are more pathogenic than others. Go et al. amplified DNA located between repetitive DNA sequences (REP-PCR). Computer-assisted cluster analysis of REP-PCR fingerprints showed two large clusters of isolates, one associated with simple gastritis and the other one with duodenal ulcer (77). Preliminary results suggest that there may also be a similar cluster for strains associated with gastric cancer. This is an area of research that promises to produce exciting results in the future. The continued use of PCR technology will undoubtedly shed new light on the transmission and pathogenicity of H. pylori.

REFERENCES

1. W. Krienitz, Dtsch. Med. Wochenschr., 28, 872 (1906).
2. A. Luger and H. Neuberger, Zeitschrift Für Klinische Medizin, 92, 54-75 (1921).
3. J.L. Doenges, Arch. Pathol., 27, 469-477 (1939).
4. A.S. Freedberg and L.E. Barron, Am. J. Dig. Dis., 7, 443-445 (1940).
5. E.D. Palmer, Gastroenterology, 27, 218-220 (1954).
6. B.J. Marshall, Campylobacter pylori in Gastritis and Peptic Ulcer Disease, M.J. Blaser, ed. Igaku-Shoin, New York,
7. J.R. Warren and B. Marshall, Lancet, 1, 1273-1275 (1983).
8. E. Madan, J. Kemp, T.U. Westblom, M. Subik, S. Sexton and J. Cook, Am. J. Clin. Pathol., 90, 450-453 (1988).
9. R.M. Genta, G.O. Robason and D.Y. Graham, Hum. Pathol., 25, 221-226 (1994).
10. T.U. Westblom, Helicobacter pylori in peptic ulceration and gastritis, B.J. Marshall, R.W. McCallum and R.L. Guerrant, eds. Blackwell Scientific, Boston, (1991) pp. 81-91.
11. T.U. Westblom, E. Madan, J. Kemp and M.A. Subik, J. Clin. Microbiol., 26, 1393-1394 (1988).
12. R.P. Logan, R.J. Polson, J.J. Misiewicz, G. Rao, N.Q. Karim, D. Newell, P. Johnson, J. Wadsworth, M.M. Walker and J.H. Baron, Gut, 32, 1461-1464 (1991).
13. B.J. Marshall, M.W. Plankey, S.R. Hoffman, C.L. Boyd, K.R. Dye, H.F. Frierson,Jr., R.L. Guerrant and R.W. McCallum, Am. J. Gastroenterol., 86, 438-445 (1991).

14. J. Weil, G.D. Bell, K. Powell, A. Morden, G. Harrison, P.W. Gant, P.H. Jones and J.E. Trowell, Aliment. Pharmacol. Ther., 5, 309-313 (1991).

15. J. Weil, G.D. Bell, P.H. Jones, P. Gant, J.E. Trowell and G. Harrison, Lancet, 2, 1245 (1988).

16. N.J. Talley, D.G. Newell, J.E. Ormand, H.A. Carpenter, W.R. Wilson, A.R. Zinsmeister, G.I. Perez-Perez and M.J. Blaser, J. Clin. Microbiol., 29, 1635-1639 (1991).

17. T.U. Westblom, L.M. Lagging, B.R. Midkiff and S.J. Czinn, Diagn. Microbiol. Infect. Dis., 16, 317-320 (1993).

18. T.U. Westblom, E. Madan, S. Gudipati, B.R. Midkiff and S.J. Czinn, J. Clin. Microbiol., 30, 96-98 (1992).

19. T.U. Westblom, S. Phadnis, P. Yang and S.J. Czinn, Clin. Infect. Dis., 16, 367-371 (1993).

20. G.H. Doran, D.N. Dickel, W.E. Ballinger, O.F. Agee, P.J. Laipis and N.N. Hauswirth, Nature, 323, 803-806 (1986).

21. D.H. Pershing, J. Clin. Microbiol., 29, 1281-1285 (1991).

22. P.A. Wright and D. Wynford-Thomas, J. Pathol., 162, 99-117 (1990).

23. S. Hoshina, S.M. Kahn, W. Jiang, P.H. Green, H.C. Neu, N. Chin, M. Morotomi, P. LoGerfo and I.B. Weinstein, Diagn. Microbiol. Infect. Dis., 13, 473-479 (1990).

24. S.A. Ho, J.A. Hoyle, F.A. Lewis, A.D. Secker, D. Cross, N.P. Mapstone, M.F. Dixon, J.I. Wyatt, D.S. Tompkins, G.R. Taylor and et al, J. Clin. Microbiol., 29, 2543-2549 (1991).

25. L. Engstrand, A.M. Nguyen, D.Y. Graham and F.A. el-Zaatari, J. Clin. Microbiol., 30, 2295-2301 (1992).

26. N.P. Mapstone, D.A. Lynch, F.A. Lewis, A.T. Axon, D.S. Tompkins, M.F. Dixon and P. Quirke, J. Clin. Pathol., 46, 540-543 (1993).

27. B. Morera-Brenes, R. Sierra, R. Barrantes, J. Jonasson and C.E. Nord, Eur. J. Clin. Microbiol. Infect. Dis., 13, 253-257 (1994).

28. J. Wahlfors, J.H. Meurman, J. Toskala, A. Korhonen, P. Alakuijala, E. Janatuinen, U.M. Karkkkainen, P. Nuutinen and J. Janne, Eur J Clin Microbiol Infect Dis, 14, 780-786 (1995).

29. C. Clayton, K. Kleanthous and S. Tabaqchali, J. Clin. Pathol., 44, 515-516 (1991).

30. J.T. Wang, J.T. Lin, J.C. Sheu, J.C. Yang, D.S. Chen and T.H. Wang, Eur. J. Clin. Microbiol. Infect. Dis., 12, 367-371 (1993).
31. J. Bickley, R.J. Owen, A.G. Fraser and R.E. Pounder, J. Med. Microbiol., 39, 338-344 (1993).
32. T.T. Lin, C.T. Yeh, C.S. Wu and Y.F. Liaw, Dig. Dis. Sci., 40, 2214-2219 (1995).
33. A.P. Lage, E. Godfroid, A. Fauconnier, A. Burette, J.P. Butzler, A. Bollen and Y. Glupczynski, J Clin Microbiol, 33, 2752-2756 (1995).
34. M. Ashton-Key, T.C. Diss and P.G. Isaacson, J Clin Pathol, 49, 107-111 (1996).
35. O. Kawamata, H. Yoshida, K. Hirota, A. Yoshida, R. Kawaguchi, Y. Shiratori and M. Omata, Biochem. Biophys. Res. Commun., 219, 266-272 (1996).
36. M. Hammar, T. Tyszkiewicz, T. Wadström and P.W. O'Toole, J. Clin. Microbiol., 30, 54-58 (1992).
37. P.W. O'Toole, S.M. Logan, M. Kostrzynska, T. Wadström and T.J. Trust, J. Bacteriol., 173, 505-513 (1991).
38. J.L. Valentine, R.R. Arthur, H.L. Mobley and J.D. Dick, J. Clin. Microbiol., 29, 689-695 (1991).
39. F.A. el-Zaatari, A.M. Nguyen, R.M. Genta, P.D. Klein and D.Y. Graham, Dig. Dis. Sci., 40, 109-113 (1995).
40. T. Shimada, K. Ogura, S. Ota and A. Terano, J Clin Gastroenterol, 20 Suppl 2, S100-S103 (1995).
41. J.C. Thijs, A.A. van Zwet, W. Moolenaar, M.J. Wolfhagen and J. ten Bokkel Huinink, Am. J. Gastroenterol., 91, 93-97 (1996).
42. R.L. Mueller, H.J. Hagel, G. Geymeier and W. Domschke, Wien. Med. Wochenschr., 136, 209-212 (1986).
43. M. Delmee, J.C. Debongnie, P. Warzee and P. Mainguet, Gastroenterol. Clin. Biol., 11, 550-553 (1987).
44. J.C. Debongnie, M. Delmee and P. Mainguet, Acta Gastroenterol. Belg., 52, 3-8 (1989).
45. L.G. Bazhenov, B.A. Saidkhanov and N.U. Khodzhaeva, Lab. Delo, 71-73 (1990).
46. D. Dionisio, C. Buonamici, D. Mazzotta, L. Ricciarelli, P. Pecile, M. Gabbrielli, G. Poggiali and R. Zanchi, Boll. Ist. Sieroter. Milan., 68, 197-198 (1989).

47. S. Krajden, M. Fuksa, J. Anderson, J. Kempston, A. Boccia, C. Petrea, C. Babida, M. Karmali and J.L. Penner, J. Clin. Microbiol., 27, 1397-1398 (1989).
48. D.A. Ferguson,Jr., C. Li, N.R. Patel, W.R. Mayberry, D.S. Chi and E. Thomas, J. Clin. Microbiol., 31, 2802-2804 (1993).
49. M. Song, Chung Hua Liu Hsing Ping Hsueh Tsa Chih, 14, 237-240 (1993).
50. G.A. Markova, R.K. Boshnakov, P.K. Petrov, K.V. Katsarov and Gintsburg, Molekuliarnaia Genetika, Mikrobiologia, i Virusologia, 10-15 (1994).
51. C. Li, P.R. Musich, T. Ha, D.A. Ferguson,Jr., N.R. Patel, D.S. Chi and E. Thomas, J Clin Pathol, 48, 662-666 (1995).
52. P.G. Hardo, A. Tugnait, F. Hassan, D.A. Lynch, A.P. West, N.P. Mapstone, P. Quirke, D.M. Chalmers, M.J. Kowolik and A.T. Axon, Gut, 37, 44-46 (1995).
53. A.A. van Zwet, J.C. Thijs, A.M. Kooistra-Smid, J. Schirm and J.A. Snijder, J. Clin. Microbiol., 32, 1346-1348 (1994).
54. H. Enroth and L. Engstrand, J Clin Microbiol, 33, 2162-2165 (1995).
55. N.P. Mapstone, D.A. Lynch, F.A. Lewis, A.T. Axon, D.S. Tompkins, M.F. Dixon and P. Quirke, Lancet, 341, 447 (1993).
56. S.M. Kelly, M.C. Pitcher, S.M. Farmery and G.R. Gibson, Gastroenterology, 107, 1671-1674 (1994).
57. J.E. Thomas, G.R. Gibson, M.K. Darboe, A. Dale and L.T. Weaver, Lancet, 340, 1194-1195 (1992).
58. P. Sahay, A.P. West, P.M. Hawkey and A.T. Axon, J Infect, 30 , 262-263 (1995).
59. P.D. Klein, D.Y. Graham, A. Gaillour, A.R. Opekun and E.O. Smith, Lancet, 337, 1503-1506 (1991).
60. R.J. Hopkins, P.A. Vial, C. Ferreccio, J. Ovalle, P. Prado, V. Sotomayor, R.G. Russell, S.S. Wasserman and J.G. Morris, Jr., J. Infect. Dis., 168, 222-226 (1993).
61. T.U. Westblom, S.B. Fritz, S. Phadnis, B.R. Midkiff, R. Leon-Barua, S. Recavarren, A. Ramirez-Ramos and R.H. Gilman, Acta Gastroenterol. Belg., 56 Suppl, 84 (1993).(Abstract)
62. R.J. Owen, J. Bickley, A. Hurtado, A. Fraser and R.E. Pounder, J. Clin. Microbiol., 32, 1203-1210 (1994).
63. H.X. Xia, H.J. Windle, D.G. Marshall, C.J. Smyth, C.T. Keane and C.A. O'Morain, Gut, 37, 30-34 (1995).

64. K. Schutze, E. Hentschel, B. Dragosics and A.M. Hirschl, Gut, 36, 831-833 (1995).

65. H.M. Mitchell, S.L. Hazell, T. Kolesnikow, J. Mitchell and D. Frommer, Infect Immun, 64, 1166-1172 (1996).

66. S.I. Majewski and C.S. Goodwin, J. Infect. Dis., 157, 465-471 (1988).

67. N. Akopyanz, N.O. Bukanov, T.U. Westblom, S. Kresovich and D.E. Berg, Nucleic Acids Res., 20, 5137-5142 (1992).

68. M.J. Torres and J.C. Palomares, Enferm. Infecc. Microbiol. Clin., 10, 345-348 (1992).

69. A.M. Hirschl, M. Richter, A. Makristathis, P.M. Pruckl, B. Willinger, K. Schutze and M.L. Rotter, J. Infect. Dis., 170, 473-475 (1994).

70. N. Akopyanz, N.O. Bukanov, T.U. Westblom and D.E. Berg, Nucleic Acids Res., 20, 6221-6225 (1992).

71. C.R. Lopez, R.J. Owen and M. Desai, FEMS Microbiol. Lett., 110, 37-43 (1993).

72. S. Fujimoto, B. Marshall and M.J. Blaser, J. Clin. Microbiol., 32, 331-334 (1994).

73. E.S. Drazek, A. Dubois and R.K. Holmes, J. Clin. Microbiol., 32, 1799-1804 (1994).

74. A. Hurtado and R.J. Owen, FEMS Immunology & Medical Microbiology, 8, 307-313 (1994).

75. A. Ito, T. Fujioka, T. Kubota and S. Nasu, Nippon Shokakibyo Gakkai Zasshi, 92, 185 (1995).

76. A. Hurtado, R.J. Owen and M. Desai, Res. Microbiol., 145, 585-594 (1994).

77. M.F. Go, K.Y. Chan, J. Versalovic, T. Koeuth, D.Y. Graham and J.R. Lupski, Scand J Gastroenterol, 30, 640-646 (1995).

IMMUNODIAGNOSIS OF SCHISTOSOMIASIS

Victor C.W. Tsang and Patricia P. Wilkins

Immunology Branch, Parasitic Diseases Division, National Center for Infectious Diseases, Centers for Disease Control and Prevention, U.S. Public Health Service, Atlanta, Georgia 30341-3724

ABSTRACT

The most efficacious and practical means of diagnosing human schistosomiasis is based on the detection of infection-specific antibodies. Because of their high sensitivity and specificity, antibody assays remain the most practical assays for epidemiologic studies and patient management. Antibody assays are particularly useful in the diagnosis of schistosomiasis in visitors from developed countries to endemic areas. These patients are often lightly infected, and tests that depend on detection of ova or circulating antigens are not reliable for these type of light and acute infections. Initial screening may be performed in the field or laboratory with the FAST-ELISA, using adult microsomal antigens. Species-specific confirmation is obtained by immunoblots with the same antigens.

INTRODUCTION

Human schistosomiasis, predominantly caused by the three trematodes, *Schistosoma mansoni, S. japonicum,* and *S. haematobium*, affects some 200 million people and countless livestock [7]. Despite control efforts in countries such as Egypt and the People's Republic of China, the disease continues to increase. Traditional diagnosis was based on detection of parasite eggs in stool and urine. However, the sensitivity of corporal examination is highly limited by the sporadic fecundity of the adult worms and the low rate of recovery of parasitic ova from patient samples. Therefore, patient management based solely on the presence of ova is overly conservative and may result in patients with low egg counts being undiagnosed and able to continue to transmit the disease [7]. Antibody (Ab) assays with the proper antigens (Ag) are highly specific and sensitive [7 - 9, 16, 18], but because of the persistence of antibodies, they may not be able to differentiate between new and old infections, or to distinguish successful treatments from failures. Given this consideration, the ideal assay will probably be an Ag detecting test. The presence of adult Ags should imply an active infection state. However, despite considerable advances in this field, Ag detecting assays still lack specificity and sensitivity.

Various diagnostic methods for schistosomiasis were reviewed recently by Maddison [7]. A tabulation of the major types of assays and brief comments on test characteristics are shown in table 1.

IMMUNODIAGNOSTIC METHODS

Our approach to routine reference diagnosis of schistosomiasis is first to screen all sera with the Falcon assay screening test-enzyme-linked immunosorbent assay (FAST-ELISA) [5] and then to confirm and speciate with the enzyme-linked immunoelectrotransfer blot (EITB) [1, 20]. The FAST-ELISA and confirmatory EITB both derive their sensitivity and specificity from microsomal Ags of adult worms [16, 18, 21]. Methods for producing Ags and immunoblot strips will not be detailed in this review. The reader is referred to previous publications for these procedures [16, 21].

I. FAST-ELISA

A. Procedure:

1. FAST™ beads, molded onto sticks and attached to 96 well lids are incubated with diluted microsomal antigens (2 mg protein in 0.05 M Tris/KOH, 0.3 M KCl, 2.0 mM EDTA, pH 8.00) for a minimum of 1 hr at room temperature (RT). Gentle agitation on a rotatory shaker during this sensitization period is important. The polypropylene baffled tray supplied with the FAST kits is used for this step, and 19 ml (2 ml for the small trough and 17 ml for the large) of Ag solution is needed for each lid. Polystyrene containers should be avoided, since the styrene will compete with the sticks to bind the Ags. Sticks are first washed by being sprayed for 20 sec with PBS/Tween (phosphate buffered saline, 0.10 M NaCl, 0.05 M Na_2HPO_4/NaH_2PO_4, pH 7.20 with 0.3% v/v Tween-20) from a plastic, hand-pumped, garden sprayer and then rinsed with d.H_2O. Sensitized sticks may be air dried and stored desiccated for at least 1 year at RT with little decrease in Ag activity.
2. Calibrated reference standard serum samples containing 0, 10 - 1000 U/μl of specific Ab activity are included on every plate. Three μl of each standard and unknown serum is diluted with 97 μl of PBS/Tween into individual wells of a 96-well FAST plate. Construction of the reference standard serum series has been previously discussed [5].

TABLE 1. Summary of major assays for the diagnosis of schistosomiasis.

Ref	Method	Antigen	PROS	CONS
			Antibody detection assays:	
26	IFA	AW	Good sensitivity (85% with egg+ cases)	No relationship between titer and onset or severity
11	IFA	GASP/CAA	Light infection detectable	No differences between acute/chronic
11	RIA	PSAP	Correlation with egg counts acute patients predominantly IgM	Radioactive isotopes
12	RIP	AW	Acute serum has bands: 55,52,35kd	Requires technical expertise and radioactive isotopes
10*	COPT	EGG	80% sensitive/97% specific	Low level infection not detectable
	ELISA	SEA	83% sensitive/94% specific	Low level infection not detectable
	ELISA	CEF6	92% sensitive/95% specific	Not species specific, no correlation w/egg count
9	ELISA	MAMA	Quantitative, 96% sensitive/99% specific	
6	WB	Sm31/32	99% Sensitivity/99% specific	Requires technical expertise
			Antigen detection assays:	
3,4	ELISA	CAA(serum)	Correlation with egg count detects <1 ng/ml serum	60-70% Sensitivity with 1-100 eggs/g
24	ELISA	CAA(urine)	Correlation with egg count	Present in urine 6 weeks post Rx

* Adult worm antigens were better than egg antigens for detecting low level infections (<100 eggs/g); egg antigens typically detect only 65-85% (7 assays tested). AW assays (n=14) ranged from 55-91%.

3. Just prior to assay, antigen-coated beads are sprayed with PBS-Tw, followed by H_2O. Excess water is then shaken off.
4. After beads are inserted into the serum plate, with standards in the top row of each plate, they are incubated while being shaken with a Mini-orbital shaker, setting 7-8, (Bellco, Vineland, NJ) for 5 min, at RT.
5. Beads are washed as in step 1.
6. Beads are inserted into trough plate containing a diluted enzyme-conjugated anti-human reagent. We use peroxidase (POD; donor: H_2O_2 oxidoreductase, EC 1.11.1.7) conjugated, immunoaffinity purified goat anti-human antibodies prepared in our laboratory [17]. The molar ratio of peroxidase to IgG for the goat anti-human conjugate (GAHG-POD) is 3.9:1, and that of goat anti-mouse (GAMs-POD) is 3.1:1. The optimum concentration of conjugate to use is that which will saturate all bound human Igs within 5 min. Each lot of conjugate reagent should be titrated to determine this optimal working

concentration. Beads are incubated while being shaken, (Bellco shaker setting 5-6) for 5 min at RT.

7. Beads are washed as in step 1.
8. Beads are inserted into substrate plate, and incubated with shaking, (Bellco shaker setting 7-8) for 2 min, at RT. Our preferred substrate (150 μl/well) is the 3,3',5,5' tetramethylbenzidine (TMB) peroxidase substrate system from Kirkegaard & Perry, Inc., Gaithersburg, MD. This very stable substrate system, employing two ready-to-mix (1:1 ratio) solutions, is convenient to use and reliable under field conditions.
9. Absorbance is measured at 650 nm ($\Delta A_{650\,nm}$/2 min) with an ELISA plate reader. It is also possible to estimate FAST-ELISA results by eye under field conditions [13].

B. Data reduction:

1. The Ab-dependent increase of $\Delta A_{650\,nm}$/2 min can be described by a 4-parameter line equation:

$$y = \frac{(a - d)}{1 + (x/c)^b} + d$$

where:

$y = \Delta A_{650\,nm}$/2 min

x = Ab activity in units (U/μl of serum)

a = zero Ab concentration response

d = infinite Ab concentration response

b & c are derived constants describing the slope of the curve.

2. ELISA reader-controlling software such as the Softmax™ (Molecular Devices Corp, Palo Alto, CA) readily employs this mode of data manipulation for immunochemical reactions. Digital data of raw absorbance value are directly processed into a standard curve from which Ab activities of unknown serum samples can be directly derived. Figure 1 shows a typical standard curve. The shape of the curve is sigmoidal and covers a reaction range of 2 - 4 $\log_{10}$. The top and bottom of this curve covering the extremely low (⇒ a) and high (⇒ d) reaction ranges, respectively, are rather flat. Consequently, reaction values extrapolated from or near these ranges are subjected to a degree of uncertainty. The greatest accuracy and confidence is associated with values derived from the middle portion of this curve, where the slope is near linear. This high-confidence range generally encompasses 1.5 Log_{10}. Our original assay volume is 3 μl per serum; however, this assay can be performed with 1 - 10 μl of serum. Therefore, it is possible to conduct the assay in

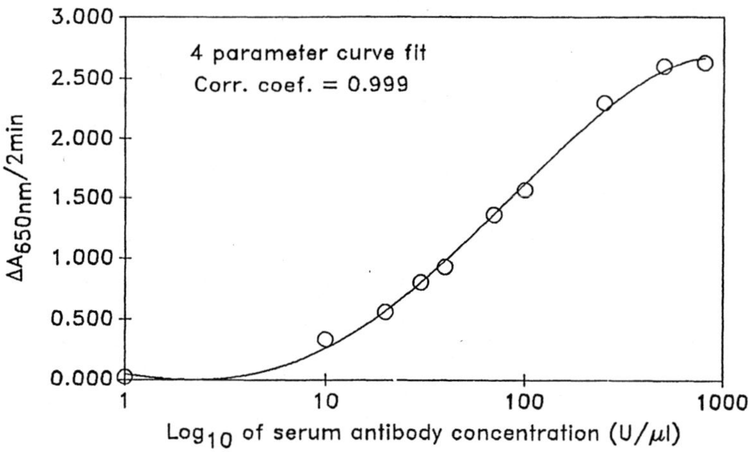

Figure 1
A typical FAST-ELISA standard curve generated by the ELISA reader and controlling software (Softmax™) of Molecular Devices Corp, CA. The Ab-dependent increase of $\Delta A_{650\ nm}/2$ min from each standard was directly plotted against its Ab concentration. Concentrations of Ab for unknown samples were directly derived from this line equation.

a manner where all unknowns fall within the high-confidence range. As long as the calibration reference serum series is also assayed at the same volume, all units derived from the standard curve will be correctly reflected. The limitation for using a larger volume of sera is that samples with high Ab concentrations will exceed the 3.5 absorbance readability limit of most readers. In this situation, the sera must first be diluted (e.g. 1:10) before assay in order to obtain proper quantitation.

C. Data interpretation and special test characteristics:

The FAST-ELISA is an internally calibrated test which quantifies Ab. Data from this test are on a continuous scale; no titers are provided. The advantage of such assays is their ability to measure Ab levels over time. The internal calibration derived from a fixed set of standard reference sera also minimizes test-to-test errors. The continuous data scale also allows a quantitative approach for the determination of a "cut-off" value. From our original study of 234 parasitologically confirmed serum samples [5], 10 U/μl was the reaction level which separates infected from uninfected patients as well as those who were infected with heterologous pathogens. This level was set for the *S. mansoni* microsomal Ag (MAMA) system and referenced against our primary standard SMPR-P. In a recent study with an equivalent Ag from *S. japonicum* (JAMA) and a different infection serum standard (SJPRC-1), the "cut-off" value was set much lower, at approximately 1.5 U/ml.

All serum Ab levels are calibrated against a specific standard reference serum pool and are only meaningful when referenced against the standard. Therefore, the unit designation is relative. Since all serum Ab quantifications and test performances are dependent on the reference standard sera, it is critical that this reagent be carefully prepared. We select only sera from parasitologically confirmed patients with the maximum diversity in terms of Ag recognition (assayed by EITB with crude Ag). We also include sera with high and low quantitative levels of Ab in the pool. Once the primary reference serum is established, it is carefully frozen (-85 #C) in small (~0.2 ml) aliquots. Initial unit designation for the primary standard was set at 100 U/ml. Secondary and tertiary working serum standards are formed in the same manner; however, all subsequent reference serum pools must be referenced against the primary standard with respect to their Ab content (U/ml). A different primary reference serum pool is created for each disease.

D. Test performance:

The determination of a "cut-off" point is largely an exercise in statistics. The ideal "cut-off" is the level above which all serum samples from confirmed infected individuals react, and below which lie all uninfected or heterologously infected individuals. This point is rather artificial, since it is difficult to expect an infection with one single worm to cause a clear increase in Ab production, such that the signal produced far exceeds that of the background noise. In reality, the concept of the "cut-off" point works much better than expected, at least in the FAST-ELISA with MAMA as Ag. The original study describing the FAST-ELISA, based on 234 serum samples, and a previous study, based on 511 serum samples describing MAMA purification [16, 21] and the k-ELISA [22, 23], both showed that it is possible to set a "cut-off" level which produced a test sensitivity and specificity of 99% (Figure 2). In other words only 1% of the confirmed infected samples reacted below the "cut-off" and 1% of the non-exposed negative controls reacted above the "cut-off" level.

E. Field-applicable versions of the FAST-ELISA:

Diagnostic assays designed for the well-equipped laboratory in developed countries tend to be dependent on sophisticated equipment and reagents, requirements that are difficult to fulfill in developing areas of the world. In particular, immunoassays of tropical diseases should ideally be field applicable. To this end, we have modified the basic FAST-ELISA and

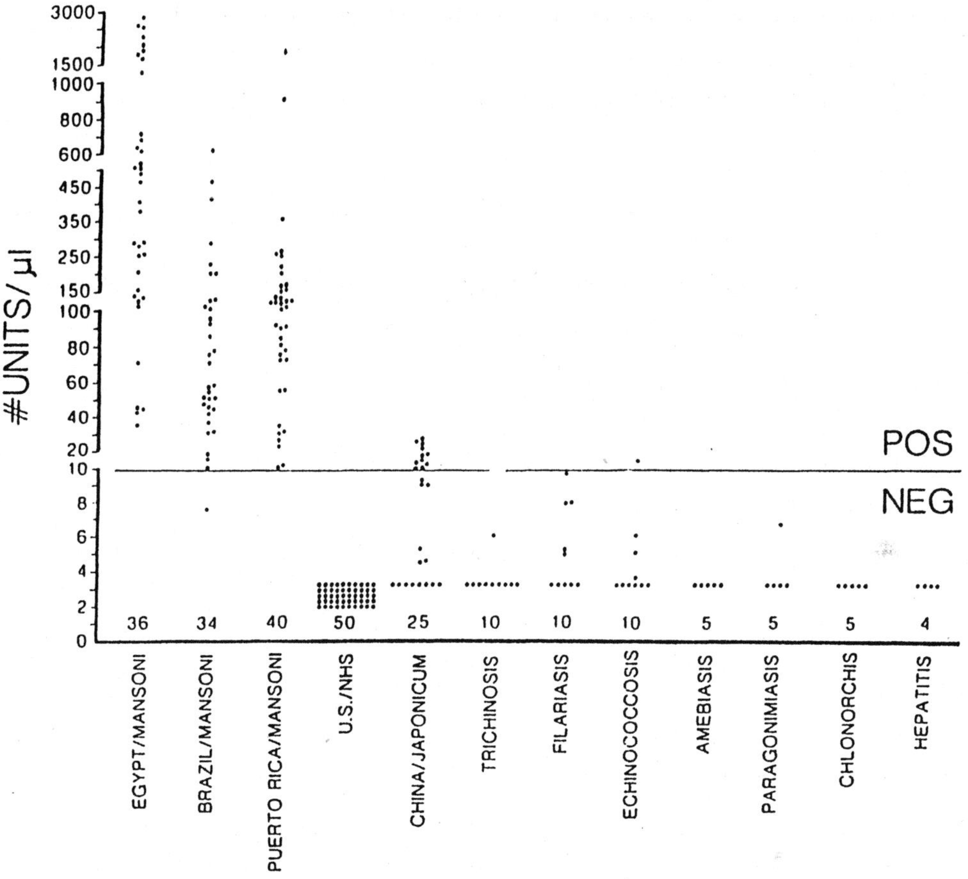

Figure 2
FAST-ELISA results (U/ml, y-axis) from 234 individuals. Groups of serum samples are identified on the x-axis. "Cut-off" point was set at 10 U/ml. At this level the FAST-ELISA for *S. mansoni* has a species sensitivity and specificity of 99%. (reproduced from Hancock and Tsang [5]).

developed two field-applicable versions of this assay [14]. These field-applicable assays are functional under minimally equipped field conditions, yet they retain the rapid, simple, and reliable qualities of the original FAST-ELISA. In addition, both assays can function with plasma, sera, or whole blood (fresh & lysed). Test specificity and sensitivity for these are same that of the FAST-ELISA.

II. EITB (Western Blot)

The EITB for all three human species of schistosomiasis is performed with microsomal Ag from adult worms. For the purpose of this review, the method of production for Ags and immunoblot strips will not be detailed. The reader is referred to previous publications for these procedures [16, 21].

A. Procedure:

1. Ag strips, which are usually stored frozen, are pre-wetted and incubated with PBS/Tw for approximately 5 min.
2. Test sera are diluted 1:50 with PBS/Tw/milk (5.0% nonfat dried milk in PBS/Tw) directly into Accutran™ incubation trays (Schleicher and Schuell, Keene, NH). Total volume of diluted sample used per trough is 0.5 ml. One strip is added (blotted surface up) to each sample and the tray gently rocked back and forth to ensure total submersion of the strip. The usual incubation time with sera is 1 hr, with constant rocking. Sera may be incubated with strips overnight, but the serum dilution should be 1:100. The strips must not be allowed to dry or emerge above fluid surfaces during incubation, otherwise severe "water marks" may form.
3. The Accutran™ washer and warm (56°C) PBS/Tween are used to remove unbound serum components. The usual wash consists of four 5-min washes.
4. Strips are incubated for 1 hr with anti-Ig-POD. The conjugates used for this assay should have at least a 1:2 (IgG:POD) molar ratio [17]. Using conjugates with lower antibody-to-enzyme ratios lessens the sensitivity of this test.
5. Strips are washed again (five times with PBS/Tween and twice with PBS alone). Bound Abs are visualized with DAB substrate solution [19].

B. Data interpretation and special test characteristics:

Figure 3 exemplifies typical EITB results for all three human schistosome infections. The specific reactions that constitute positive reactions are GP30 for *S. mansoni*; GP18, G23, or GP29 for *S. japonicum*; and GP23 for *S. haematobium* [1]. It is important to note that these specific reaction bands occur only on strips containing the respective species-specific microsomal Ags. Thus, reactions at the 30KD position on HAMA strips do not constitute a positive reaction. It is also vital to recognize that each of the species-specific diagnostic bands has a characteristic appearance of its own, and this quality should be considered part of the recognition criteria. For example, the GP30 of *S. mansoni* is a typically diffused glycoprotein band which usually appears as a doublet, although reactivity to only one member of the doublet is common. Therefore, a control serum that reacts to both doublets of GP30 should be included in every run. Another example is the specific bands for *S. japonicum*; these

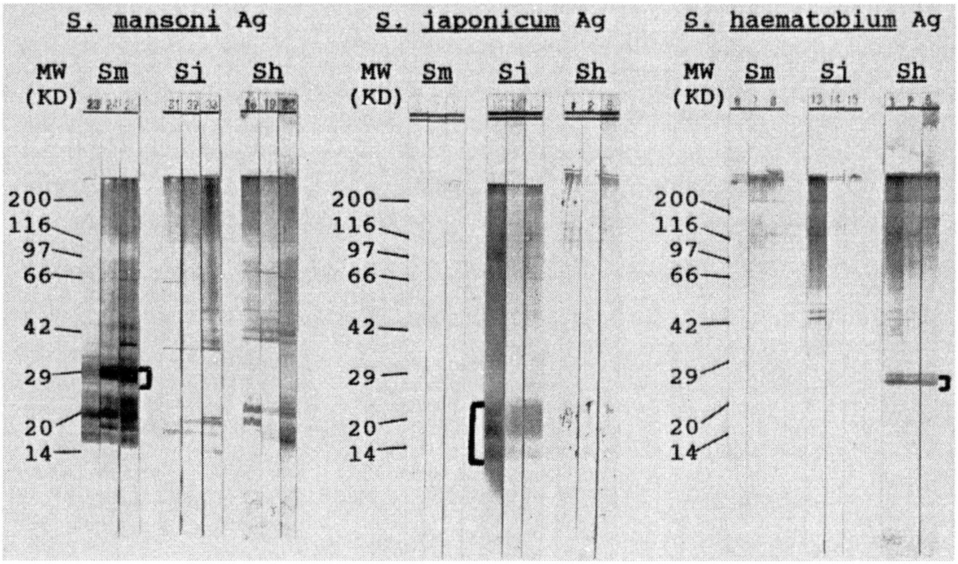

Figure 3
Typical EITB results for serum samples from three human *Schistosoma* infections. The specific reactions (marked by brackets) that constitutes positive reactions are: GP30 for *S. mansoni*; GP18, G23, or GP29 for *S. japonicum*; and GP23 for *S. haematobium*. The same serum samples were tested with each of the three kinds of Ag strips.

usually appear as three distinct groups of glycoproteins, each composed of multiple small bands, giving them a faintly striped appearance. Again, not all sera will recognize all three groups. Consequently, a control that reacts to all three groups of Ag should be included. In addition to the species-specific diagnostic bands, other reactions are common on all three Ag strips. However, all irrelevant bands are ignored during data interpretation.

C. Test performance:

Since the first description of MAMA in 1983 [21], these Ags have been used extensively in research and routine reference diagnosis. We have not encountered a single parasitologically proven false positive. Thus, the specificity of the EITB with MAMA for the diagnosis of *S. mansoni* infections is 100%. The only possible exception to this observation involve the reaction to GP30 from the serum of a U.S. serviceman stationed in Panama. We have not been able to associate this patient with any previous risk of exposure to *S. mansoni*, yet his serology and symptoms are consistent with schistosomiasis.

Table 2 illustrates the sensitivity and specificity of the EITB with HAMA antigens from *S. haematobium* in detecting species-specific infections. Similar specificity and sensitivity was found with EITB using MAMA and JAMA antigens for their respective infections.

D. Modification of the EITB:

The original version of the EITB generally requires a minimum of 3 hr and as long as overnight to complete. For field applications and for rapid screening, we have developed a rapid version of this assay. This rapid blot [2] requires approximately 15 min to perform and possesses the same sensitivity and specificity as the original version of the EITB.

III. Cost estimate and commercial availability

FAST-ELISA was estimated to cost $0.13 per serum sample tested in our laboratory. This cost assumes triplicate assays for each serum sample and calibration standards to occupy the first row of every plate. This estimate is based on running one to two plates per week. Larger test numbers decrease dramatically the cost per test.

Actual cost of the EITB is estimated at $3.75 per serum sample, of which $0.65 is accounted for by Ag and strip production costs [15].

Commercial costs for these assays are higher than the above estimates. The sources for these assays are as follows:

1. Microbiology Reference Laboratory
   10700 Walker Street
   Cypress, California 90630-4737
   Telephone (800) 445-0185
2. Specialty Laboratories, Inc.
   P.O. Box 92722
   Los Angeles, California 90404-3900
   Telephone (800) 421-7110

DISCUSSION

The specificity and, to a lesser extent, the sensitivity of immunological assays, are critically dependent on Ag characteristics and purity. The FAST-ELISA and confirmatory EITB both derive their sensitivity and specificity from microsomal Ags of adult worms. The microsomal Ags from the adult worms of all three *Schistosoma* spp. contain large amounts of specific components

TABLE 2: Sensitivity and specificity of the EITB with HAMA*

Disease/infective agent**	Donor Origin	EITB+	EITB-	Total
S. haematobium	Luxor, Egypt	93	2	95
S. haematobium	Sudan	44	2	46
S. haematobium	Malawi	11	0	11
S. haematobium	Africa	7	0	7
S. haematobium	Angola	1	0	1
S. haematobium	Ghana	11	0	11
S. haematobium	Kenya	1	0	1
S. haematobium	Liberia	1	0	1
S. haematobium	Madagascar	1	0	1
S. haematobium	Mali	2	0	2
S. haematobium	Namibia	1	0	1
S. haematobium	Nigeria	1	0	1
S. haematobium	Rep. of S. Africa	1	0	1
S. haematobium	Saudi Arabia	2	0	2
S. haematobium	Sierra Leone	3	0	3
S. haematobium	Upper Volta, Egypt	0	1	1
S. haematobium	Yemen	0	1	1
Total *S. haematobium*		**180**	**6**	**186**
S. mansoni	Brazil	0	16	16
S. mansoni	Puerto Rico	0	10	10
S. japonicum	P.R. China	0	8	8
Paragonimiasis	P.R. China	0	10	10
Trichinosis	U.S.A.	0	8	8
Cysticercosis	Peru	0	8	8
Cysticercosis	Mexico	0	10	10
Echinococcus granulosis	Peru	0	15	15
Echinococcus multiloccularis	Alaska, U.S.A.	0	11	11
Amebiasis	Mexico	0	8	8
Hepatitis A	U.S.A.	0	5	5
Hepatitis B	U.S.A.	0	8	8
Hepatitis C	U.S.A.	0	5	5
Clonorchiasis	P.R. China	0	3	3
Filariasis	India	0	4	4
Onchocerca volvulus	Guatemala	0	8	8
Normal Human Sera	USA, non-travelers	0	20	20
Total heterologous infections + normal		**0**	**157**	**157**

* Based on the 186 samples from parasitologically confirmed *S. haematobium* infected donors, the sensitivity of the EITB with HAMA is 97% (180/186). The specificity of this test is 100%, since none of the 157 heterologous infection and/or normal donor sample reacted positive.

** Serum or plasma samples diluted 1:50 with PBS/Tw were assayed by EITB with HAMA as antigen (see Material and methods). Disease/infective agents were confirmed by direct observation of infective agents, larva, or ova in donor tissues, urine, or feces. Heterologous and normal humans serum samples were derived from donors without exposure risk to *S. haematobium*, on the basis of their geographical segregation and travel history. All heterologous infection serum donors, except those from U.S.A., are socioeconomically similar to the *S. haematobium* donors.

and minimal quantities of nonspecific or irrelevant materials. Therefore, the specific activity of these reagents is high. Antigens with high specific activity are especially critical for assays such as the FAST-ELISA that use crude Ags. In this type of assay, all antigen components are adsorbed on a solid matrix such as polystyrene. Binding of Ab onto any component, whether specific or nonspecific, will produce a reaction. In contrast, in the EITB all components of the antigen mixture are separated from each other by SDS-PAGE (sodium dodecyl sulfate polyacrylamide gel electrophoresis), and reactions to specific and nonspecific antigen bands can be distinguished from each other by the differences in their relative molecular weights.

High-resolution confirmatory assays, especially immunoblot-based assays are generally difficult and expensive. The EITB is no exception; therefore, screening with this assay is impractical. However, screening with a rapid simple assay such as the FAST-ELISA and confirming all positives with the EITB is efficacious. This approach is identical to that used in HIV antibody testing and has performed well in our reference diagnostic service for schistosomiasis.

The issue of Ab persisting even after treatment remains a deterrent for choosing Ab-detection assays over Ag-detection tests for the diagnosis of schistosomiasis. However, a recent study in our laboratory shows that the assumption of persistent Ab after successful treatment may not be universally correct [25]. In our study [25], nine rhesus monkeys were successfully treated with praziquantel, five others were not treated, and one was treated but not cured (Figure 4). The quantitative IgG1 ratios (pretreatment divided by posttreatment absorbance values) of the nine cured monkeys at 40 weeks posttreatment were generally higher than those of the five that were not treated and the one that was not cured (Figure 4). In this laboratory study, we confirmed the infection status of all animals by postmortal perfusion. In humans the success or failure of treatment cannot be confirmed. Therefore, persistence of Ab after treatment may also be the result of incomplete cure. An Ag-detecting assay with high sensitivity and specificity will be the ultimate tool to help resolve this issue. Until such an assay is developed, the screening and confirmatory approach for detecting Abs will remain practical. The utility of this diagnostic approach is clearly illustrated in the following study [27].

In 1992 two U.S. Peace Corps volunteers (PCVs) developed central nervous system schistosomiasis due to infection with *Schistosoma haematobium* following recreational water exposure at Cape Maclear on Lake Malawi, an African lake considered by many to be free of schistosomiasis. To determine the transmission potential and risk for acquiring schistosomiasis in Lake Malawi, a cross-sectional survey of resident expatriates and visitors to Malawi was done during March and April 1993. Blood specimens were collected to determine the seroprevalence of

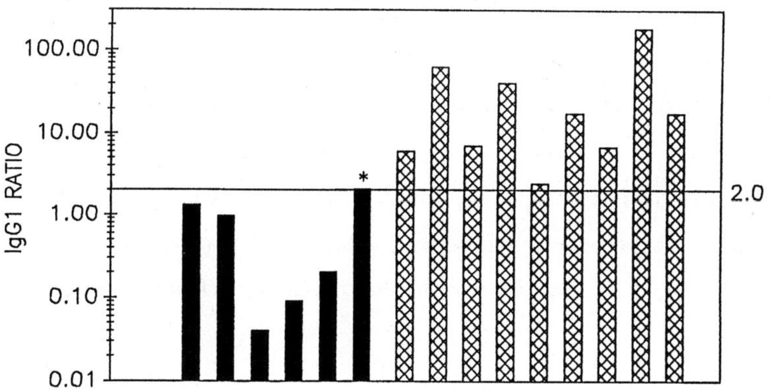

Figure 4
Quantitative, Ag-specific IgG1 ratios (FAST-ELISA) in treated and untreated rhesus monkeys at 40 weeks post-therapy. Nine monkeys were successfully treated with praziquantel, 13 weeks post-infection (cross-hatched bars). Five others were not treated (solid bars), and one was treated but not cured (*). IgG1 ratios were derived by dividing individual pretreatment values by posttreatment values.

S. haematobium and *S. mansoni* using FAST-ELISA and immunoblot analysis. A survey for vector snails was conducted along Lake Malawi's southwestern shore.

The study population of 955 expatriates included 99 PCVs, 55 U.S. State Department employees, 151 other U.S. citizens and 650 non-U.S. foreign nationals. One-third of the study population (305/955) had serologic evidence of current or past schistosome infection. The seroprevalence was 33% (141/440) among expatriates whose freshwater exposure was limited to Lake Malawi; *S. haematobium* antibodies were found in 135/141 (96%) seropositive specimens. Risk of seropositivity increased with the number of freshwater exposures at Lake Malawi resorts. While many resort areas in the southwestern Lake region posed a significant risk, Cape Maclear was the location most strongly associated with seropositivity (OR 2.9; 95% CI 1.6-5.1); the etiologic fraction attributable to freshwater exposure at Cape Maclear was 56%. *Bulinus globosus*, the snail vector of *S. haematobium* in Malawi, was found along the lake shore.

S. haematobium infection is highly prevalent among expatriates and tourists in Malawi, some of whom have suffered serious CNS complications. Recreational water contact at popular resorts on Lake Malawi is the most likely source of infection. Epidemiologic and environmental investigations indicate that transmission of schistosomiasis is occurring in Lake Malawi, a previously unrecognized site of transmission.

ACKNOWLEDGMENT

The bulk of this article has been published in Clinics in Laboratory Medicine 11:1029, 1991 by the same authors, and is reproduced here under the Copyrights Act, a "work of the United State Government" for which copyright protection under Title 17 of the United State Code is not available. The use of trade names in this report is for identification only and does not constitute endorsement by the Public Health Service or by the United States Department of Health and Human Services.

REFERENCES

1. Andrews J, Tsang VCW, Maddison SE. J Inf Dis 160:171, (1989).
2. Brand JA, Tsang VCW: Immunoassay 10:237, (1989).
3. Deelder AM, de Jonge N, Boerman OC, et al. Am J Trop Med Hyg 40:268, (1989).
4. de Jonge N, Gryseels B, Hilbreath GW, et al. Trans Roy Soc Trop Med Hyg 82:591, (1982).
5. Hancock K, Tsang VCW. J Immunol Met 92:167, (1986).
6. Idris MA, Ruppel A. J Helminthol 62:95, (1988).
7. Maddison SE. Diagn Microbiol Infect Dis 7:93, (1987).
8. Maddison SE. In: Helminthic Diseases (Immunodiagnosis of Parasitic Diseases, vol 1). Walls KW, Schantz PM (eds) Orlando, Academic Press, (1986) p 1
9. Maddison SE, Slemenda SB, Tsang VCW, et al. Am J Trop Med Hyg 34:484, (1985).
10. Mott KE, Dixon H. Bull World Health Organization 60:729, (1982).
11. Nash TE. Ann Intern Med 97:740, (1982).
12. Norden AP, Strand M. Am J Trop Med Hyg 34:495, (1985).
13. Richards FO, Tsang VCW, Brand JA, Hancock K. Trop Med Parasitol 40:332, (1989).
14. Rossi CL, Tsang VCW, Pilcher JB. J Clin Microbiol 29:000, (1991).
15. Tsang VCW, Brand JA, Boyer AE. J Inf Dis 159:50, (1989).
16. Tsang VCW, Hancock K, Kelly MA, et al. J Immunol 130:1366, (1983).
17. Tsang VCW, Hancock K, Maddison SE. J Immunol Met 70:91, (1984)
18. Tsang VCW, Hancock K, Maddison SE, et al. J Immunol 132:2607, (1984).
19. Tsang VCW, Hancock K, Wilson M, et al. CDC training course manual (monograph),p 26, (1986).
20. Tsang VCW, Peralta JM, Simons AR. Methods in Enzymol 92:377, (1983).
21. Tsang VCW, Tsang KR, Hancock K, et al. J Immunol 130:1359, (1983).
22. Tsang VCW, Wilson BC, Maddison SE. Clin Chem 26:1255, (1980).
23. Tsang VCW, Wilson BC, Peralta JM. Methods in Enzymol. 92:391, (1983).
24. Van Leishout L, de Jonge N, Bassily S, et al. Am J Trop Med Hyg 44:323, (1991).
25. Wilkins PP, Slemenda SB, Maddison SE, et al. Am J Trop Med Hyg 45:629, (1991).
26. Wilson M, Sulzer AJ, Walls KW. Am J Trop Med Hyg 23: 1072, (1974).
27. The Morbidity and Mortality weekly report, CDC 42:565, (1993).

PART III

VIRAL INFECTIONS

THE EVOLUTION OF HANTAVIRUSES

Xinyang Zhao and John Hay
Department of Microbiology
State University of New York at Buffalo
Buffalo,NY 14214

ABSTRACT

Hantaviruses exist in most regions of the world. The many different strains identified thus far have widely divergent roles in human disease and infect a wide range of rodent hosts. The sequence data available for the genomes of these viruses allows us to study indirectly the evolutionary patterns of the *hantaviruses*. In this paper, we describe relationships among the M genomic segments of *hantaviruses*, and attempt to relate these to the evolutionary relationships of the virus' rodent hosts.

RESULTS and DISCUSSION

The study of virus evolution is difficult, for one very obvious reason - we have no fossils to use in tracing where and when a virus might have originated or how it may have changed. What we can do is speculate on possible virus evolution by comparing genomic sequences of currently existing strains from different parts of the world. The *hantaviruses* represent a good basis for this kind of study, since many strains exist worldwide in a variety of different animal hosts, and cause a variety of different diseases in man.

Hantaviruses belong to the family Bunyaviridae, a group of viruses with three segmented, negative polarity, single-stranded RNA genomes. Presently, the family is divided into five genera: *phlebovirus, bunyavirus, nairovirus, hantavirus and tospovirus*. The *tospoviruses* are plant viruses, and all the other genera, except *hantaviruses,* use both vertebrates and invertebrates as vectors and hosts. *Hantaviruses* have rodents as their only "natural" hosts, and humans are infected tangentially as they come into respiratory contact with rodent excretions and secretions. In this respect, *hantaviruses* behave more like members of the Arenaviridae family.

One of the major diseases of man caused by *hantavirus* infection is hemorrhagic fever with renal syndrome (HFRS); this is probably an old disease, although the WHO officially labelled it as one of the newcomers in human pathology as recently as 1983 (1). Western physicians and scientists first encountered hantaviral infections as a serious problem during the Korean War, where the United Nations lost numbers of troops to a "mystery" disease involving hemorrhage and renal failure (it was HFRS). However, the etiologic agent of these HFRS cases was not discovered until 1978, when Dr. Ho Kwan Lee isolated a virus from the striped field mouse, *Apodemus agrarius*, that reacted with sera from HFRS patients. Shortly afterwards, the virus was adapted to grow in tissue culture. To date, human *hantavirus*-associated disease has been reported in 33 countries and antibodies to *hantaviruses* have been detected in 37 countries (2). All human diseases, ranging from severe to mild, were reported from Asia and Europe and, until recently, no human disease was reported in Africa (3), North America (4) and South America (5), despite the finding that *hantaviruses* were enzootic in these areas. However, the situation changed in 1993, when a fatal respiratory disease suddenly appeared in the Four Corners region of the southwestern USA. This disease was found to be caused by hitherto unknown strains of *hantaviruses*, and caused disease quite unlike any previously associated with these viruses. This new disease was called Hantavirus Pulmonary Syndrome (HPS), and was subsequently found all over North America, from Florida to Texas and Canada, although its incidence is low.

Thanks to the power of RT-PCR technology, and to the extensive molecular research on *hantaviruses*, the genomes of many viral strains have been sequenced and this has led to the development of an understanding of the phylogeny of these viruses.

Conserved regions among the M segments of seven representative viruses.

In our first study, we used the MACAW program from NIH to search for conserved and non-conserved sequences among the M genomic segments of several *hantavirus* strains (6). This genomic segment codes only for the two viral glycoproteins, G1 and G2, that are required for viral recognition of cells to be infected; these clearly play a vital role in the viral life cycle.

As in all other living creatures, viral mutation is the force behind evolution, and RNA viruses, lacking the replicative proofreading ability of DNA-based organisms, are generally subject to high mutation rates. Scanning the overall *hantavirus* M segment sequences, it is clear that the sequences of the G1 and G2 proteins from different viruses are quite diverse; even Sin Nombre virus and El Moro Canyon virus (two strains from the HPS outbreak in the southwestern USA) are only 70% homologous in G1 and G2, as well as being divergent serologically (7). With respect to the other viruses we will examine, Hantaan virus is related to Seoul virus (both Korean strains), while Puumala (from Scandinavia) and Prospect Hill virus (from Maryland) are also related (8, 9, 10). Most of the new HPS strains are closely related and are similar to Prospect Hill virus, although this last virus does not seem to be the cause of any human disease.

Seven strains of *hantaviruses* were examined, as representatives of different general groups, according to a phylogenetic tree generated by others using the PAUP and CGC pileup programs (11); these were: Hantaan, Seoul, Thailand, Pulmonary, Prospect Hill, El Moro Canyon and Puumala viruses. The Gibbs sampling strategy was used to search for blocks of sequence, 20 amino acids long as the maximum and 3 amino acids long as the minimum (17). A general examination of the distribution of these conserved regions (figure 1) shows that most of the M segment sequences are not conserved, based on the criteria used in the MACAW program. This implies that only small areas of the viral glycoproteins have to conform to a precise structural pattern. As will be seen, only some of the conserved regions are currently associated with a known function, and this analysis therefore allows the definition of additional sequences, particularly in G2, that should be examined experimentally to determine their actual importance and structural role.

We have recently analysed the M segment sequence using two separate programs that define membrane-associated regions (14). These data show: for G1, a signal sequence from aa1 to 23, a possible transmembrane domain from aa198 to 218, and two transmembrane domains spanning aa459 to 511; for G2, a "signal sequence from aa621 to 645 and a transmembrane domain from aa1097 to 1117.

The first conserved block in the M segment is from aa61 to 67, near the G1 N-terminus, and falls into a conserved linear epitope, currently mapped from aa59 to 88 (12). Interestingly, so far, the G1 monoclonal antibodies only can recognize truncated (linear) epitopes, while no G2 monoclonal sees a linear epitope (13).

From aa417 to 499, there are three conserved blocks, corresponding to a transmembrane domain in G1, and its associated sequences. As we have recently shown using mutation and laser scanning confocal microscopy, this region of the molecule is necessary and sufficient for the Golgi-targeting of G1 and, through structural association, of G2 (14). Given that the Golgi-targeting ability of hantaviruses is likely to be an essential part of the virus' life cycle, for example, as part of its persistent infection ability in rodents, we should not be surprised that this region is conserved. Particularly well-conserved within this otherwise strongly hydrophobic region are several hydrophilic residues (e.g. serines, threonines and histidines), as well as cysteine residues, and this implies an important role for these residues in Golgi-targeting, perhaps as elements of intra- and intermolecular network formation within the membrane.

The next conserved block is between aa535 and 550, corresponding to the presumed cytoplasmic tail of G1, which may have a function in recognition of the viral nucleocapsids during the maturation process. Thus, G1 appears to have only three conserved regions, one presumably externally exposed (the epitope), one involving a transmembrane region, and the third of currently unknown function.

For G2, there are three short and four longer conserved regions (figure 1). The short regions (four, four and six amino acids respectively) may represent exposed regions, and could

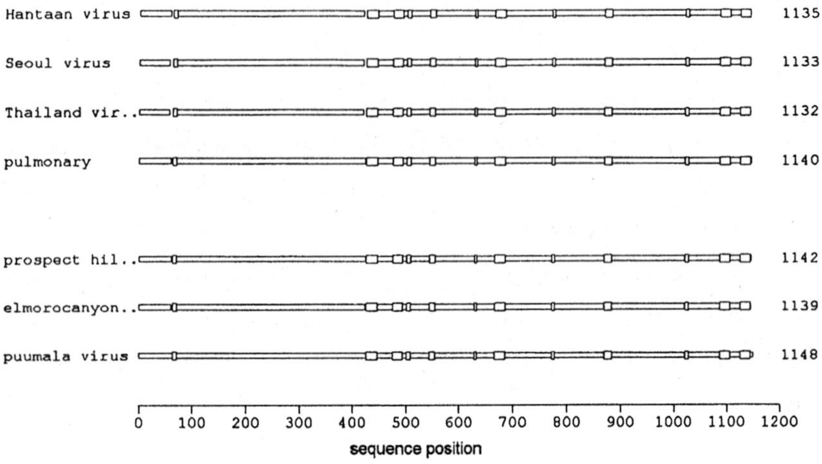

FIGURE 1

Amino acid sequence comparisons among selected *hantavirus* genomic M segments. Conserved regions, based on the MACAW program from NIH, are shown as open blocks.

also be involved in contacts between G2 and G1 that are important for tertiary structure generation. The first two longer conserved regions (closer to the amino terminus) may also represent exposed regions and may be epitopes for B or T-cell activity. The putative "signal peptide" for G2 (from aa621 to 645) is, surprisingly, not well-conserved; in fact, for Pulmonary, El Moro Canyon, Prospect Hill and Puumala viruses, there is no methionine residue present to act as a potential initiation signal, suggesting that there is no general mechanism for internal initiation of translation in *hantaviruses* that would generate G2.

The first large block of conserved sequence within G2 is from aa657 to 676, immediately following the "signal peptide" sequence, and the second is roughly in the middle of the protein. Both of these sequences have substantial hydrophilic character and may be involved in the generation of intra- or intermolecular sructure.

Like the G1 protein, the G2 protein carboxy terminus is more conserved than the amino terminus, and contains the remaining two large conserved regions. This may be partly due to the existence of a transmembrane domain in G2, which is close to the carboxy terminus. Just amino terminal to this region is a serine residue (aa1124), mapped by one group (15) as involved in fusion of the cellular membrane. While the proposed transmembrane region of G2 is not precisely picked up as conserved by the MACAW program, it has several structural features that are consistent from strain to strain, as well as conserved hyrdophilic residues reminiscient of the G1 transmembrane domain. The G2 transmembrane domain also contains the ER targeting signal

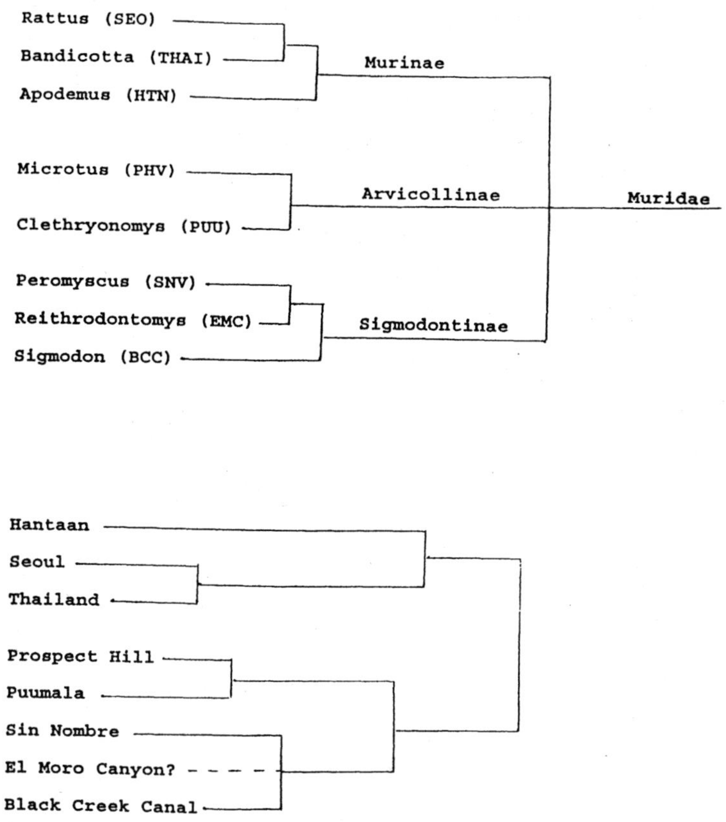

FIGURE 2

Phylogenetic trees of *hantaviruses* and of their rodent hosts. The upper panel shows the rodent tree and the lower the virus tree. The abbreviations in brackets beside the rodent names refer to the associated *hantavirus*. Neither vertical nor horizontal distances are to scale.

for the protein, which presumably resides in the highly positively-charged region close to the molecular terminus. Although there are no exact residue similarities in this putative ER targeting region, the various lysine-rich motifs in the different virus strains may represent the important ER elements (19).

Relationships between the viruses and their hosts.

Since Homo sapiens is the only apparent opportunistic host for *hantaviruses*, it is not surprising that different virus strains cause quite different diseases in man. On the other hand, we can speculate from these data that *hantaviruses* are very old viruses, residing in different rodent species for considerable lengths of time. Indeed it is possible that these viruses were resident in rodents before they evolved into the species we know today, with their quite different ecological habitats. Because of this, perhaps genomic reassortment did not play a major role in hantavirus evolution; thus far, there is good evidence for only one case of reassortment, and whether it is important for epidemiology remains to be seen (16, 17). Actual intermolecular recombination (as opposed to reassortment) is also potentially involved in evolution, but there is no evidence that this mechanism functions in these viruses. Based on these arguments, it would be interesting to compare the phylogenetic trees of *hantaviruses* with those of their rodent hosts, as has been done with papillomaviruses and human races. In figure 2, we show the phylogenetic tree of eight *hantaviruses* (20, 22, 23) alongside the phylogenetic tree of the rodent hosts (21). There is a striking similarity between the two trees, although the *muridae* have three main limbs compared to two for the viruses. Among the US viruses, it is not clear how El Moro Canyon virus is related to Black Creek Canal virus, although both of these have about 75-80% homology in the glycoprotein region with Sin Nombre virus. Thus the likelihood that these viruses are of ancient origin and have evolved with their rodent hosts seems strong.

Finally, the base composition of viruses can also be a useful guide to their origin and evolution. *Hantaviruses* have a C+G content of about 40%, close to the C+G content of human (and mammalian) DNAs, while larger viruses, such as herpes simplex viruses, have much more divergent C+G content. For *hantaviruses*, which need to use the mRNA caps of the host cell and do not shut off host RNA synthesis, it would make sense to have a genome that is characteristic of the host genome. This also implies that these small RNA viruses may have been derived from or co-evolved with host cell sequences at some point in their evolution.

ACKNOWLEDGEMENTS

This work was supported by grants from the USPHS (AI25376) and by a Pilot Grant from SUNY at Buffalo.

REFERENCES

1. WHO. Bull WHO, 61, 269-275 (1983).

2. M.P. Chumakov, I.N. Gavrilovskaya and Y.A. Myasnikov. Arch. Virol. S1, 49-56 (1990).

3. J.F. Saluzzo, J.P. Digoutte and A.P. Adam. Trans. Roy. Soc. Trop. Med. Hyg. 79, 874-875 (1985).

4. J.W. LeDuc, G.A. Smith and L.R. Bagley. New Eng. J. Med. 307, 624-626 (1982).

5. J.W. LeDuc, G.A. Smith and F.P. Pinheiro. Am. J. Trop. Med. Hyg. 34, 810-815 (1985).

6. G.D. Schuler, S.F. Altschul and D.J. Lipman. Prot. Str. Fn. Gen. 9, 180-190 (1991).

7. N. Torrez-Martinez, W. Song and B. Hjelle. Virology, 211, 336-338 (1995).

8. J. Arikawa, H.F. Lapentiere, L. Iacono-Commors, M. Wang and C.S. Schmaljohn. Virology, 176, 114-125 (1990).

9. M.A. Parrington, P.W. Lee and C.Y. Kang. J. Gen. Virol. 72, 1845-1854 (1991).

10. S.Y. Xaio, J.W. LeDuc, Y.K. Chu and C.S. Schmaljohn. Virology, 198, 205-217 (1994).

11. C.F. Spiropoulou, S. Morzunov, H. Feldmann, A. Sanchez, C.J. Peters and S.T. Nichol, Virology, 200, 715-723 (1994).

12. B. Hjelle, F. Chavez-Giles, N. Torrez-Martinez, T. Yamada, J. Sarisky, M. Ascher and S. Jennison. J. Gen. Virol. 75, 2881-2888 (1994).

13. M.Wang, D.G. Pennock, K.W. Spik and C.S. Schmaljohn. Virology, 197, 757-766 (1993).

14. X. Zhao and J. Hay. J. Virol., submitted (1996).

15. Y. Isegawa, O. Tanishita, S. Ueda and K. Yamanishi. J. Gen. Virol. 75, 3273-3278 (1994).

16. W.W. Henderson, M.C. Monroe, S.C. St. Jeor, W.P. Thayer, J.E. Rowe, C.J. Peters, J.E. Peters and S.T. Nichol. Virology 240, 602-610 (1995).

17. D. LI., A.L. Schmaljohn, K. Anderson and C.S. Schmaljohn. Virology, 206, 973-983 (1995).

18. C.E. Lawrence., S.F. Altschul, M.S. Boguski, J.S. Liu., A.F. Neuwald and J.C. Wootton. Science, 262, 208-214 (1990).

19. M.R. Jackson, T. Nillson and P.A. Peterson. EMBO J., 9, 3153-3162 (1990).

20. B. Hjelle, S. Jenison, N. Torrez-Martinez, T. Yamada, K. Notle, R. Zumwalt, K. MacInnes and G. Myers. J. Virol., 68, 592-596 (1994).

21. M.D. Carlton and G.G. Mussa in Orders and Families of Recent mammals of the World, S. anderson and J.K. Jones, eds, Texas Tech University Press (1984).

22. N. Torrez-Martinez, W, Song and B. Hjelle. Virology, 211, 336-338 (1995).

23. E.V. Reskov, P.E. Rollins, T.G. Ksaizak, C.J. Peters and S.T. Nichol. Virology, 210, 482-489 (1995).

LABORATORY DIAGNOSIS OF HEPATITIS C

Frederick S. Nolte, Ph.D.
Department of Pathology and Laboratory Medicine
Emory University School of Medicine
Atlanta, Georgia 30322

ABSTRACT

In 1989 the hepatitis C virus was cloned, some 20 years after the first suggestion that hepatitis C virus(es) (HCV) existed. Since that time there has been rapid development of serological tests for detection of antibody to HCV and molecular tests for detection, quantitation, and characterization of HCV RNA. The development and performance characteristics of these test methods are reviewed and their implications for diagnosis, prognosis, and monitoring patients on therapy are discussed.

DISCOVERY

For more than twenty years researchers sought the virus or viruses that were thought to be the cause of non-A, non-B (NANB) hepatitis. Attempts to identify a virus by conventional immunological and virological methods failed (1). In 1989, a group from Chiron Corp., in conjunction with the CDC cloned a portion of the hepatitis C virus (HCV) genome from the blood of a chimpanzee inoculated with serum of a patient with NANB hepatitis (2). The HCV is now recognized as the major cause of what used to be known as NANB hepatitis.

VIROLOGY

HCV has a single-stranded 9.4 kb positive-sense RNA genome. It contains one large open reading frame capable of encoding a polyprotein precursor of 3011 amino acids(3). HCV is most closely related to the pestiviruses (e.g., bovine viral diarrhea virus, and hog cholera virus), and flaviviruses (e.g., yellow fever, Dengue, and Japanese encephalitis viruses). HCV is thought to be a distinct genus in the *Flaviviridae* family. The virus has proven difficult to cultivate and has not yet been visualized by electron microscopy.

The structural proteins are encoded at the 5' end of the genome. The 5' non-coding region (NCR) of 341 nucleotide precedes the large coding sequence and is the most highly conserved region of the genome. It may contain an internal ribosome entry site and play an important regulatory role during viral replication. The genome encodes three structural proteins; an unglycosylated, basic core protein, and two putative envelope glycoproteins designated E1 and E2. Four non-structural (NS) domains follow; NS2 to NS5. The NS2 protein is extremely hydrophobic and its function has not been defined. NS3 protein contains a viral protease involved in polyprotein processing, and a putative RNA helicase involved in unwinding the genome for replication. The NS4 region is extremely hydrophobic, varies considerable between types, and has no known function. The NS5 region encodes a RNA-dependent RNA polymerase that replicates the genome (4).

Variable regions throughout the HCV genome have been identified including the NS3, NS4, NS5, core, E1, and the 5' NCR. Based on phylogenetic analysis of the NS5 and E1 nt sequences from samples around the world, there are 6 major genotypes (5). The major types are further divided in a more closely related subtypes. Genotype is designated with an Arabic number and subtype by a lower case letter. Genotypes 1, 2 and 3 have broad distribution, whereas, the others have more restrictive geographic distributions. Subtypes 1a and 1b are common in the U.S. Patients with HCV type 1 have lower response rates to interferon therapy than those infected with HCV types 2 and 3. The genome plasticity is also demonstrated in individual patients, where the virus exists simultaneously as a series of related, but immunologically distinct, variants termed "quasispecies"(6). Although mutations occur throughout the genome, most of the mutations leading to escape variants occur in a hypervariable region of E2/NS1.

DISEASE

Prior to 1990, ≈250,000 new cases of hepatitis C occurred annually in the U.S. About 25% of infected individuals become ill with hepatitis. Among patients in long term follow-up, at least 70% have persistent or intermittent ALT elevations. In those with ALT elevations, liver biopsy almost always confirms the presence of chronic inflammatory change. HCV infection can lead to cirrhosis and hepatocellular carcinoma. End-stage chronic hepatitis C is now one of the leading indications for liver transplantation. The frequencies of these serious consequences of HCV infection are still being defined (4).

HCV is transmitted through exposure to blood (7). Blood transfusion, hemodialysis intravenous drug abuse, and occupational exposure to blood are risk factors. Transmission to sexual and family contacts is uncommon, as is perinatal transmission from infected mother to child. Many infected individuals have no identifiable risk factors.

Interferon alpha is the only agent of proven efficacy in the treatment of hepatitis C (8,9) Standard treatment is 3 million units, three times per week for 6 months. Nearly all patients relapse and require retreatment. A sustained improvement was seen in only 10 to 51% of patients enrolled in treatment trials. HCV genotype and viral load seem to influence response to interferon (10,11) .

SEROLOGICAL TESTS

Shortly after the discovery of HCV, an EIA was developed to detect antibody to a cloned peptide (c110-c) (12,13). The second and third generation EIAs include various mixtures of recombinant antigens, and detect a broader range of anti-HCV antibodies (14). The presence of antibody to HCV is confirmed by the use of supplemental assays. The most common supplemental assay is a recombinant immunoblot assay (RIBA) (15). The recombinant antigens included in each successive generation of RIBA (1 to 3) have been changed to improve the performance characteristics of the test (16). These serological tests are important tools for screening blood donors and for diagnosis of HCV infection in symptomatic patients. To date, there are no commercially available tests for IgM anti-HCV antibody, or for HCV antigens.

False-positive EIA results may be due to nonspecific antibody interactions or cross reacting antibodies. False-negative EIA results may be due to selection of the capture antigens, window of seronegativity, fluctuating antibody titer, immunosuppresion, or HCV genotype. HCV antibody tests do not distinguish between current and previous disease.

HCV RNA DETECTION

Indications for HCV RNA testing include: suspected acute hepatitis C prior to seroconversion; suspected chronic hepatitis C in seronegative patients; resolution of indeterminate RIBAs; monitor patients on interferon therapy; evaluation of occupational exposure from a high-risk source; and virological safety testing of plasma products. Reverse transcription-polymerase chain reaction (RT-PCR) has been used extensively to demonstrate HCV RNA in serum or plasma (17). Early attempts were made with nested primer RT-PCR (18). Nested RT-PCR greatly increases the amount of product DNA and obviates the need for hybridization with an oligonucleotide probe, but the sequential PCRs require transfer of reagents which increases the likelihood of false-positives due to product carry-over contamination. Single primer pair RT-PCR followed by hybridization with a probe has similar analytical sensitivity but with fewer opportunities for cross-contamination. It has become clear that sample integrity, method of RNA isolation, and primer selection are all important determinants in the overall performance of RT-PCR assays for detection of HCV RNA (19,20).

A commercially available single enzyme RT-PCR (AMPLICOR HCV Test, Roche Molecular Systems) represents a significant advance and should help standardize HCV RNA detection (21). A DNA polymerase from *Thermus thermophilus*, in the presence of manganese carries out both the reverse transcription and DNA polymerase functions. Use of a single enzyme to carry out both functions has several distinct advantages. The RT-PCR can now be done in a single tube which reduces the potential for contamination. Since the *T. thermophilus* enzyme is heat-stable, the RT step can be done at a higher temperature that increases the specificity of primer extension and is compatible with uracil-N-glycosylase protocol for carry-over prevention. The AMPLICOR HCV Test has compared favorably with "home-brewed" RT-PCR assays in two clinical evaluations (22,23).

Although the HCV RNA assays complement the available serological tests, the technology is not without limitations (24). The viremia in chronic HCV infections may be intermittent, so a single negative RT-PCR does not rule out infection. The technology is expensive and not currently feasible for large scale screening. There is no standardized method for detection of HCV RNA. Finally, proficiency testing programs in Europe (Eurohep) and the U.S. (CAP) indicate that many laboratories have serious problems distinguishing positive from negative samples (25).

HCV RNA QUANTITATION

The quantitation of HCV RNA in serum may be important as a prognostic indicator, to assess virological response to interferon therapy, and to monitor patients receiving therapy. A variety of methods have been used to quantitate HCV RNA including, endpoint dilution RT-PCR (26), competitive RT-PCR (27), multicyclic RT-PCR (28), nucleic acid sequence based amplification (29), RT-PCR with a single internal quantitation standard (30), and branched DNA (bDNA) (31). Truly quantitative RT-PCR-based assays are difficult to construct because the efficiency and kinetics of amplification may vary between specimens. Since the total amplification of the target molecules is described by the equation $(1+x)^n$ where x is the average per cycle efficiency and n is the number of cycles, it is easy to understand how minor variations between samples can be exponentially compounded. In contrast, bDNA is essentially a series of simple hybridization reactions involving capture, target, amplifier, and label probes that does not depend upon amplification of the target molecules and, as a result, is inherently more quantitative but less sensitive than RT-PCR based methods.

Two assays RT-PCR with a single internal quantitation standard (HCV MONITOR, Roche Molecular Systems), and bDNA (Quantiplex HCV, Chiron) are currently available as research use only kits. Much of the data documenting response to interferon in clinical trials has been collected with the Quantiplex assay. Quantiplex provides quantitative results over a 3 log dymaic range (Quantiplex version 2; $2x10^5$ to $5.7x10^7$ genome equiv./ml), and is very reproducible (CVs $\approx$20%). The second version of the Quantiplex assay was released in January 1996 and provides for better quantitation of HCV genotypes 2 and 3 (32,33). Quantiplex version 1 underestimated the HCV RNA levels of type 2 by a factor of 3 and type 3 by a factor of 2. Monitor also provides quantitative results over a 3 log range ($1.0x10^3$ to $5.0x10^5$ copies /ml) but is less reproducible than Quantiplex (CV $\approx$40%). Monitor values are on average 10- to 20-fold less than Quantiplex values for the same sera (34). Although the relative accuracy of the two assays cannot be assessed without an independently quantitated "gold standard' HCV RNA preparation, the proportional systematic error appears to be greater with the MONITOR assay (34). Both MONITOR and Quantiplex are highly complex tests that require skilled operators to achieve reliable results but are feasible for routine clinical use. HCV RNA quantitation has proven to be a valuable research tool but its role in the management of individual patients needs to be clarified.

GENOTYPING AND NUCLEIC ACID SEQUENCE ANALYSIS

The effects of interferon differ among the various HCV genotypes. Genotype 1b is resistant to interferon, with rates of complete response from 10 to 40% in published trials. In contrast, complete response rates with genotypes 2a and 2b are reported to be 60 to 90%. Within genotype 1b, response to interferon is correlated with the number of mutations in the NS5A gene (35). Genotyping and nucleic acid sequence analysis may provide important information that can help predict response to therapy.

HCV genotypes also differ in their biological effects. HCV type 1b is associated with lower HCV RNA levels and a higher prevalence of cirrhosis in alcoholics than type 1a. HCV type 1a is associated with a more aggressive course of viral hepatitis following liver transplantation than type 1b. Type 2 is associated with a greater degree of inflammatory activity in the liver than in type 1. Variations in HCV genotypes have major implications for design of HCV diagnostics, therapeutics, and vaccines.

A variety of methods and regions of the genome have been used for genotype analysis. Sequence analysis from E1, NS5, core, and 5' NCR regions have been used to construct phylogenetic trees (36). For the most part there is an excellent concordance among the types and subtypes assigned based on different HCV genomic regions. The 5' NCR is so well conserved that not all subtypes are represented when it is used for phylogenetic analysis. Full or partial sequence analysis, restriction fragment length polymorphism analysis, type-specific RT-PCR, and a line probe assay have all be used to genotype HCV.

A line probe assay kit (Innogenetics) is available for research use only. It employs a RT-PCR to amplify a sequence from the 5' NCR region and a membrane with immobilized probes to detect internal variable sequences within the biotinylated PCR product. Reactive lines are visualized colorimetrically using a streptavidin-alkaline phosphatase conjugate and a substrate. The pattern of reactive lines defines the genotype. The line probe assay discriminates between the most common genotypes and demonstrated excellent concordance with reference methods in two recent evaluations (37,38).

HEPATITIS C DIAGNOSTIC TESTS: CURRENT STATUS

The practicing clinician can use the following diagnostic aids to evaluate patients with suspected hepatitis C: 1) measurement of ALT levels, 2) liver biopsy, 3) serological tests (EIA and RIBA), and 4) molecular methods (RT-PCR, bDNA, and sequence analysis) (39).

Elevated ALT levels merely defines existing liver disease and correlates poorly with HCV-related disease activity, particularly in chronic hepatitis C (40). Significant histological abnormalities may be found among HCV-infected persons with normal ALT levels (41).

Liver biopsy is important in evaluation of patients with chronic but not acute hepatitis C. Biopsies provide useful baseline measures of disease severity and also provide useful prognostic information on the potential for treatment response. The presence of cirrhosis diminishes the likelihood for successful treatment (42).

The presence of anti-HCV antibody almost always indicates true HCV infection, particularly when the ratio of sample value/cutoff value exceeds 2.5 (39). Antibody tests are usually negative in acute hepatitis C. Patients with persistently normal ALT values and EIA reactivity, particularly if they have no symptoms, require RIBA testing. Indeterminate RIBA results may be resolved using RT-PCR.

If treatment is contemplated, a baseline HCV RNA level by bDNA or quantitative RT-PCR and genotype determination are useful. Best response rates are seen in patients with lower levels and in patients infected with type 2 and 3. Mutation analysis of NS5A region in HCV type 1b may also help predict response. If ALT values do not return to normal within 1 to 3 months of treatment, the likelihood of later success is low, and treatment may be stopped. If ALT values return to normal, a second HCV RNA level should be done; and if HCV RNA is detected, the likelihood of a sustained response is low. An increased interferon dose may be considered in these patients. If the HCV RNA is not detected by bDNA or quantitative RT-PCR, a sensitive quantitative RT-PCR should be done to confirm the absence of viremia (43). If the virus still persists, even at low levels, the likelihood of success is low. Here too, an increased interferon dose might be considered. If qualitative RT-PCR is negative, the treatment will probably be successful. Successful treatment is defined as the persistent absence of HCV RNA for at least 6 months after therapy is discontinued.

REFERENCES

1. H.J. Alter. Ann Internal Med 115, 644-649 (1991).

2. Q.-L. Choo, G. Kuo, A.J. Weiner, L.R. Overby, D.W. Bradley and M. Houghton. Science 244, 359-362 (1989).

3. M. Houghton, A. Weiner, J. Han, G. Kuo and Q.L. Choo. Hepatol 14, 381-388 (1991).

4. J.A. Cuthbert. Clin Microbiol Rev 7, 505-532 (1994).

5. P. Simmonds, A. Alberti, H.J. Alter, F. Bonino, D.W. Bradley and C. Brechot et al. Hepatol 19, 1321-1324 (1994).

6. M. Martell, M.I. Esteban, J. Quer, J. Genesca, A. Weiner, R. Estaban, J. Guardia and J. Gomez. J Virol 66, 3225-3229 (1992).

7. M.J. Alter, H.S. Margolis, K. Krawczynski, F.N. Judson, A. Mares, W.J. Alexander, P.Y. Hu, J.K. Miller, M.A. Gerber, R.E. Sampliner, E.L. Meeks and M.J. Beach. N Engl J Med 327, 1899-1905 (1992).

8. A.M. Di Bisceglie, P. Martin, C. Kassiandides, M. Lisker-Melman, L. Murray, J. Waggoner, Z. Goodman, S.M. Banks and J.H. Hoofnagle. N Engl J Med 321, 1506-1510 (1989).

9. G.L. Davis, L.A. Balart, E.R. Schiff, K. Lindsay, H.C. Bodenheimer, R.P. Perrillo, I.M. Jacobson, J. Payne, J.L. Dienstag, D.H. Van Thiel, C. Tamburro, J. Lefkovitch, J. Albrecht, C. Meschievitz, T.J. Ortegao and A. Gibas and the Hepatitis Interventional Therapy Group. N Eng J Med 321, 1501-1506 (1989).

10. K. Yoshioka, S. Kakumu, T. Wakita, T. Ishikawa, Y. Itoh, M. Takayanagi, Y. Higashi, M. Shibata and T. Morishima. Hepatol 16, 293-299 (1992).

11. J.Y. Lau, G.L. Davis, J. Kniffen, K.-P. Quan, M.S. Urdea, C.S. Chan, M. Mizokami, P.D. Neuwald and J.C. Wilber. Lancet 341, 1501-1504 (1993).

12. G. Kuo, Q.-L. Choo, H.J. Alter, G.L. Gitnick, A.G. Redeker, R.H. Purcell, T. Miyamura, J.L. Dienstag, M.J. Alter, C.E. Stevens, G.E. Tegtmeier, F. Bonino, M. Columbo, W.S. Lee, C. Kuo, K. Berger, J.R. Shuster, L.R. Overby, D.W. Bradley and M. Houghton. Science 244, 362-364 (1989).

13. D.Y. Chien, Q.-L. Choo, A. Tabrizi, C. Kuo, J. McFarland, K. Berger, C. Lee, J.R. Shuster, T. Nguyen, D.L. Moyer, M. Tong, S. Suruta, M. Omata, G. Tegtmeier, H. Alter, E. Schiff, L. Jeffers, M. Houghton and G. Kuo. Proc Natl Acad Sci (USA) 89, 10011-10015 (1992).

14. C.L. van der Poel, D. Bresters, H.W. Reesink, A.A. Plailsier, W. Schaasberg, A. Leentvaar-Kuypers, Q.-L. Choo, S. Quan, A. Polito, M. Houghton, G. Kuo, P.N.R. Lelie and H.T.M. Cuypers. Vox Sang 62, 208-212 (1992).

15. H.J. Alter, G.E. Tegtmeier, B.W. Jett, S. Quan, J.W. Shih and W.L. Bayer and A. Polito. Transfusion 31, 771-776 (1991).

16. M.H. Sayers and D.R. Gretch. Tranfusion 33, 809-813 (1993).

17. T.A. Cha, J. Kolberg, B. Irvin, M. Stempien, E. Beall, M. Yano, Q.L. Choo, M. Houghton, G. Kuo, J.H. Han and M.S. Urdea. J Clin Microbiol 29, 2528-2534 (1991).

18. H. Okamoto, S. Okada, Y. Sugiyama, T. Tanaka, Y. Sugai, Y. Akahane, A. Machida, S. Mishiro, H. Yoshizawa, Y. Miyakawa and M. Mayumi. Jpn J Exp Med 60, 215-222 (1990).

19. F.S. Nolte, C. Thurmond and P.S. Mitchell. J Clin Microbiol 32, 519-520 (1994).

20. H.T.M. Cuypers, D. Bresters, I.N. Winkel, H.W. Reesink, A.J. Weiner, M. Houghton, C.L. von der Poel and P.N. Lelie. J Clin Microbiol 30, 3220-3224 (1992).

21. K.K.Y. Young, R.M. Resnick and T.W. Myers. J Clin Microbiol 31, 882-886 (1993).

22. F.S. Nolte, C. Thurmond and M.W. Fried. J Clin Microbiol 33, 1775-1778 (1995).

23. S. Zeuzem, B. Ruster and W.K. Roth. Z. Gastroenterol 32, 342-347 (1994).

24. D. Gretch, W. Lee and L. Corey. J Clin Microbiol 30, 2145-2149 (1992).

25. H.L. Zaaijer, T.M. Cuiypers, H.W. Reesink, I.N. Winkel, G. Gerken and P.N. Lelie. Lancet 341, 722-724 (1993).

26. M. Shindo, A.M. Di Bisceglie, L. Cheung, W.-K. Shih, K. Cristiano, S.M. Feinstone and J.H. Hoofnagle. Ann Intern Med 115, 700-704 (1991).

27. H. Hagiwara, N. Hayashi, E. Mita, M. Naito, A. Kasahara and H. Fusamoto. Hepatol 17, 545-550 (1993).

28. N. Ishiyama, K. Katayama, N. Ishimi, S. Takahashi, H. Igarashi, H. Nakajima, T. Andoh, S. Saito and T. Aoyagi. Japanese J Gastroenterol 6, 1396. (1992).

29. T. Kievits, B. van Gemen, D. van Strijp, R. Schukkink, M. Diruks, H. Adriansc, L. Malek, R. Sooknanan and P. Lens. J Virol Methods 35, 273-286 (1991).

30. W.K. Roth, J.-H. Jee, B. Ruster and S. Zeuzem. J Clin Microbiol 34, 261-264 (1996).

31. M.S. Urdea. Clin Chem 39, 725-726 (1993).

32. M.L. Collins, C. Zayati, J.J. Detmer, B. Daly, J.A. Kolberg, T.A. Cha, B.D. Irvine, J. Tucker and M.S. Urdea. Anal Biochem 226, 120-129 (1995).

33. J.Y.N. Lau, P. Simmonds and M.S. Urdea. Lancet 346, 425-426 (1995).

34. E. Hadziyannis, M.W. Fried and F.S. Nolte. J Clin Microbiol (in press), (1996).

35. N. Enomoto, I. Sakuma, Y. Asahina, M. Kurosaki, Murakami, C. Yamamoto, Y. Ogura, N. Izumi, F. Marumo and C. Sato. N Engl J Med 334, 77-81 (1996).

36. K.-i. Ohba, M. Mizokami, T. Ohno, K. Suzuki, E. Orito, Y. Ina, J.Y.N. Lau and T. Gojobori. Virus Res 36, 201-214 (1995).

37. A. Andonov and R.K. Chaudhary. J Clin Microbiol 33, 254-256 (1995).

38. J.Y.N. Lau, G.L. Davis, L.E. Prescott, M. G, K.L. Lindsay, K.P. Qian, M. Mizokami and P. Simmonds and the Hepatitis Intervential Therapy Group. Ann Intern Med 124, 868-876 (1996).

39. V. Tedeschi and L.B. Seeff. Ann Intern Med 123, 383-385 (1995).

40. Z.D. Goodman and K.G. Ishak. Semin Liver Dis 15, 70-81 (1995).

41. A. Alberti, G. Morsica, L. Chemello, D. Cavalletto, F. Noventa and P. Pontisso et al. Lancet 340, 697-698 (1992).

42. P. Jouet, S. Roudot-Thoraval, D. Dhumeaux and J.M. Metreau. Gastrol 106, 686-690 (1994).

43. D.R. Gretch, C. dela Rosa, R.L. Carithers Jr., R.A. Willson, B. Williams and L. Corey. Ann Intern Med 123, 321-329 (1995).

POLYMERASE CHAIN REACTION ASSAYS FOR THE DETECTION OF CYTOMEGALOVIRUS IN ORGAN AND BONE MARROW TRANSPLANT RECIPIENTS

Mary Jo Evans,[1,2,7]*, Yadira Edwards-Spring[3], Jean Myers[8], Alice Wendt[7], Deborah Povinelli[7], Daniel Amsterdam[4,5], Kate Rittenhouse-Diakun[3], Donald Armstrong[3], Brian M. Murray[4], Steven J. Greenberg[1], and Marie Riepenhoff-Talty[2,3,6,7]
Department of Neurology, Roswell Park Division[1] and Departments of Pediatrics[2], Clinical Laboratory Science[3], Medicine[4], Pathology[5] and Microbiology[6], State University of New York at Buffalo. Division of Infectious Diseases, Children's Hospital of Buffalo, NY[7], Virology Laboratory, Erie County Medical Center, Buffalo, NY[8]

ABSTRACT

Cytomegalovirus (CMV) infection is ubiquitous and results in a wide spectrum of clinical manifestations ranging from asymptomatic infection to severe life threatening disease. Infection in normal children and adults usually causes no symptoms but in the immuncompromised host, CMV may

---

Current affiliation and address for author who has moved since completion of this study: Yadira Edwards-Spring
Lederle-Praxis Biologicals
211 Bailey Road
West Henrietta, NY 14586

result in severe opportunistic infections with high morbidity and mortality. Historically, virus detection was dependent on culture of the virus or on a centrifugation culture system referred to as a shell vial assay. The shell vial assay frequently lacked sensitivity and was unable to detect infection in its early phase. Also, as with culture assays, the results were affected by antiviral therapy. The CMV antigenemia assay was developed to provide more rapid results and has gained wide usage. This assay is limited to detection of the virus in white blood cells and is more sensitive than culture or the shell vial assay.

Application of the polymerase chain reaction (PCR) to these problems has resulted in the development of assays for CMV which are more sensitive than previously available methods. This method employs liquid hybridization with ^{32}P labeled probes and gel retardation analysis for detection of amplified DNA specific for each virus.

A comparison of the detection of CMV by an antigenemia assay or the PCR method in the leukocytes of renal transplant patients revealed that the PCR assay detects cytomegalovirus earlier and more consistently than the antigenemia assay.

Finally, the application of a fluorescent dye detection system and image analysis of the acrylamide gel with a laser scanner provides additional sensitivity to the detection of cytomegalovirus, as well as avoiding the use of radioactivity, making the assay more adaptable to the clinical laboratory.

Key Words: Polymerase Chain Reaction, Antigenemia, Cytomegalovirus

INTRODUCTION

Cytomegalovirus infection in immunocompromised patients is a major cause of morbidity and mortality. Some reports indicate that as many as 90% of bone marrow and kidney transplant recipients develop CMV infections, either due to reactivation of a latent virus by immunosuppressive drugs or by transmission of the virus from a latently virus infected donor (1-3). The current growth in the number of organ and tissue transplants has markedly increased the need for rapid and reliable diagnosis of this virus (4,5).

Diagnosis of CMV may be performed by 1) conventional cell culture using human fibroblasts, 2) shell vial centrifugation culture also employing human fibroblasts as well as monoclonal antibodies and immunofluorescent staining techniques, 3) serological methods, 4) the CMV antigenemia assay which employs monoclonal antibody to detect CMV antigen in peripheral blood leukocytes, or 5) nucleic acid hybridization detection. All of these techniques have their limitations.

Conventional cell culture is sensitive, but slow. Cytopathic effects (CPE) may take 14 days or more to develop. The shell vial assay is more rapid requiring 24 - 48 hours for initial immunofluorescence detection. Additionally all culture methods may be affected by antiviral therapy.

Serological studies may not be helpful in diagnosis of an active infection, since CMV IgM antibodies may not be present in reactivation infections or in the immunocompromised host. Alternatively IgM antibodies may persist for months after the infection is resolved.

Antigenemia assays have been shown to be sensitive and specific, but they are labor intensive and can only be applied to peripheral blood leukocytes. Comparative studies have indicated that the antigenemia assay is more sensitive than either virus culture or the shell vial assay(4,5).

The polymerase chain reaction (PCR) allows for the selective amplification of DNA sequences to which oligonucleotide primers can be specifically hybridized. When applied to the diagnosis of CMV this method compared favorably with virus isolation , in situ hybridization, and southern blotting (6-14). Others have compared the antigenemia assay with PCR and the PCR has performed as efficiently or slightly more efficiently in the detection of CMV (15,18). PCR has the added advantage of applicability to specimens other than peripheral blood leukocytes.

We have compared the antigenemia assay with PCR for the diagnosis of CMV infection in renal transplant patients and have developed a fluorescent PCR which can detect cytomegalovirus without loss of sensitivity and without the necessity of a radioactive hybridization step.

MATERIALS AND METHODS

Patients and Specimens: One hundred thirteen heparin treated blood samples were obtained from 16 renal transplant patients (#1-16), five HIV positive patients (#17-21), three patients with non-CMV diagnosis (#22-24), and one patient with CMV pneumonia (#25). Antigenemia assays were performed by the Virology Laboratory at Erie County Medical Center, Buffalo, NY. PCR assays were done in the Laboratory of Molecular Diagnostics for Infectious Agents at Roswell Park Cancer Institute. Samples for PCR were transported on dry ice. All assays were blinded.

Fluorescent assays were performed on ACD (acid/citrate/dextrose) treated blood samples from bone marrow transplant recipients at Roswell Park Cancer Institute, submitted to the Molecular Diagnostics Laboratory at Children's Hospital of Buffalo for PCR screening for CMV.

Preparation of peripheral blood leukocytes (PBL): Leukocytes were separated from whole blood by sedimentation with 5% dextran as described by van der Bij, et.al. (4,5). Six mL of whole blood was mixed with 1.5 mL of 5% dextran solution in phosphate buffered saline (PBS). The blood was incubated at 37° C for 15-20 minutes at a 45° angle until the blood separated into layers. The upper layer containing the PBLs was transferred to a sterile tube and centrifuged for 10 min at 300 x g. The contaminating red cells were lysed using 5 mL of 1x red cell lysing solution (0.826% ammonium chloride, 0.1% potassium bicarbonate, 0.037% EDTA). The leukocytes were washed with 5 mL of PBS, pH 7.4. The steps were repeated if red blood cells were still present. The remaining cell pellet was suspended in 1 mL of PBS, pH 7.4 and a cell count was performed. Peripheral blood leukocytes were resuspended at a concentration of 1×10^6 cells per mL in Hank's Balanced Salt Solution (Gibco). Cell pellets containing 2×10^6 cells were prepared and frozen at -70° C for the PCR assays. Aliquots for the antigenemia assay were prepared at 1.5×10^6 cells/mL in PBS at pH 7.4.

Antigenemia Assay: Cytospin slides were prepared according to the reported procedures of van der Bij, et.al. (4,5). Briefly, slides were prepared by cytocentrifugation to contain approximately 50,000 PBLs. Two slides from each specimen were examined. Most of the PBLs (95%) were polymorphonuclear leukocytes. Slide preparations were fixed with formalin and 0.5% Nonidet P40 (United States Biochemical, Cleveland, OH) was employed for cell membrane permeabilization as described by Gerna et. al. (19). Slides were incubated with mouse monoclonal antibody mixture (IgG_1)

C10, C11 directed against the CMV lower matrix phosphoprotein (pp65) (Biotest Diagnostics Corp., Denville, NJ and Biotest AG, Dreich, Germany). After washing the slides were incubated with goat anti-mouse FITC conjugates (Biotest). Positive cells demonstrated apple green (IF) nuclear staining when viewed under 200 x magnification using the appropriate microscopy. Results were expressed quantitatively as number of positive cells per 50,000 PBLs.

DNA Extraction: High molecular weight DNA was purified from leukocytes using proteinase K extraction and ethanol precipitation (20,21). The pellet of peripheral blood leukocytes was resuspended in 200 μL of 100mM KCl, 10mM Tris, pH 8.3, 150 mM $MgCl_2$ to create a single cell suspension. The pellet was then transferred to a cryovial tube containing 200 uL of 10 mM Tris, pH 8.3, 25 mM $MgCl_2$, 1% Nonidet P40 and 1% Tween 20 to dissolve the cell membrane and inhibit DNAse and RNAse. Proteinase K (Amresco, Solon, OH) was added to a final concentration of 120 μM and the solution was incubated at 37° C overnight or at 60° C for 1 hour. The proteinase K was inactivated at 98° C for 10 min, and the DNA extract (200 μL) was transferred to a 1.5 mL microcentrifuge tube. The salt concentration was adjusted to 0.2 M NaCl and two volumes of cold absolute ethanol was added. The mixture was precipitated at -70° C for at least 1 hour and the precipitated DNA was collected by centrifugation in a microcentrifuge at 14,000 x g for 15 min. at 4° C. The pellet was air dried and redissolved in 200 μL water.

Polymerase Chain Reaction: Polymerase chain reaction was performed using immediate early antigen (IE) primers and late antigen (LA) primers corresponding to conserved regions of the CMV genome. Primer sequences are given in Table I. An additional set of primers, specific for the T cell receptor beta chain constant region gene (TcR/βc) were also used as control to assure the integrity of the DNA extracted from patients specimens and to detect inhibition of polymerization in the extract. Amplifications were performed in a total volume of 100 μL, consisting of 30 μL of DNA or control solutions in water, 1.5 mM $MgCl_2$, 10 mM Tris/Cl, pH 8.3, 50 mM KCl, 0.01% gelatin, 0.2 uM dNTPs, 0.2 uM of each primer and 1.0 unit of Taq Polymerase (Perkin-Elmer, Norwalk, CT). Samples were overlaid with two drops of mineral oil and amplified in a Perkin Elmer 480 thermal cycler. The amplification program included a 5 min denaturation step at 94° C followed by 35 cycles of 1 min at 94° C, 1 min at 55° C and 2 min at 72° C.

Control DNAs: Positive control DNA for CMV was obtained from Advanced Biotechnologies Inc., (Columbia, MD). These DNA preparations were extracted from virus preparations which were quantitated for tissue culture infectious dose, 50% ($TCID_{50}$), so that dilutions of the DNA could be used to approximate the concentration of original infectious virus. Negative control DNA was extracted from cultured Jurkat cells (NIH AIDS Research and Reference Reagents Program, contributed by Dr. Arthur Weiss) a human T cell line.

Liquid Hybridization and Gel Retardation: Radioactive detection of amplified DNA was conducted as described previously (21) Briefly, oligonucleotide probes were 5' end labeled with ^{32}P by incubating 40 pM of probe with 60 uCi of gamma ^{32}P ATP and 10 units of T_4 polynucloetide kinase for 1 hour at 37° C. Hybridization was performed in a total volume of 50 μL, containing 20 μL of PCR amplified DNA, 30 μL of a reaction mixture containing 2.5 x

TABLE I

Primer	Sequence[4]	Fragment length
CMV/LA+[1]	CAC GCA ACT TTT GGC CGC CAC ACC TGT CAC	417
CMV/LA-	CAC CAC GCA GCG GCC CTT GAT GTT T	
CMV/LA PRO	TAG GTG CTT CGG CTC ATC ACC AAC GTG GAA	
CMV/IEA+[2]	ATG AAG TGT ATT GGG CTA ACT ATG CAG AGC ATG TAT GA	360
CMV/IEA-	TTG TCT CTC TCC TCA TCC AAA ATC TTA AAT ATT TTC T	
CMV/IEA PRO	ACC AAG AAC TCA GCC TTC CCT AAG ACC ACC AAT	
TcR/βc +[3]	GTG TTC CCA CCC GAG GTC GCT GTG TTT GAG CC	124
TcR/βc-	GTG CTG ACC CCA CTG TGC ACC TCC TTC CCA TT	
TcR/βc PRO	ATC AGA AGC AGA GAT CTC CCA CAC CCA AAA GGC CAC ACT G	

[1] CMV late antigen primers and probe
[2] CMV immediate-early protein primers and probe
[3] T cell receptor Beta constant region gene primers.
[4] Primers and probes for CMV are protected by patent applications #08/428,370 & 08/600,764

10^5 cpm of labeled probe in 0.15 M NaCl. Labeled probes and PCR products were denatured at 98-100° C for 10 min and annealed at 55° C for 1 hour. Following hybridization, 20 μL of each sample was electrophoresed on a 5% acrylamide gel at 200 V for 1 hour in Tris/Borate/EDTA (TBE) buffer. The gel was placed on filter paper, packaged into a plastic bag and exposed to Kodak X-OMAT AR film (Eastman Kodak, Rochester, NY) at room temperature overnight. Film were developed using a Kodak M35A X-OMAT Processor (Eastman Kodak, Rochester, NY).

Contamination Controls: Sample preparation and analysis of amplified products were conducted in separate laboratories. All reagents were aliquoted, only pipette tips containing barriers and dedicated pipetting equipment were used and all other recommended precautions were adhered to (15). Each sample was tested three times in three different experiments and only samples giving positive results with both sets of primers and with the TcR/βc primers were considered positive.

Fluorescent Detection: Detection of amplified DNA on acrylamide gels with SYBR Green I (Molecular Probes, Inc.) using the FluorImager (Molecular Dynamics, Inc.) was as described in reference 20. Briefly, amplified DNA (20 μL) was mixed with 5 μL of loading buffer, containing 10 μg/mL Bromphenol Blue, 50% glycerol and a 1:20,000 dilution of SYBR Green I as received from the supplier. After electrophoresis, without removing the gel from the glass plates, it was inserted into the FluorImager. The area to be scanned was selected. Scan conditions were a pixel size of 200 microns, normal scan speed, digital resolution of 16 bits, one channel and PTM voltage of 500. Filter 0 and filter 1 were calibrated to 515 tLc. The scan required under three minutes and the computerized image was enhanced and labeled directly for production of a hard copy.

RESULTS

We have assayed 113 specimens from 25 renal transplant patients by an antigenemia assay and by PCR for the presence of CMV. Figure 1 is an example of the actual data produced. Each panel shows a reagent, no DNA control, six patient specimens, a negative DNA control and two concentrations of positive control CMV DNA. Patients specimens in lanes 2, 4, 5, and 7 were positive for both the immediate early protein gene, in the top

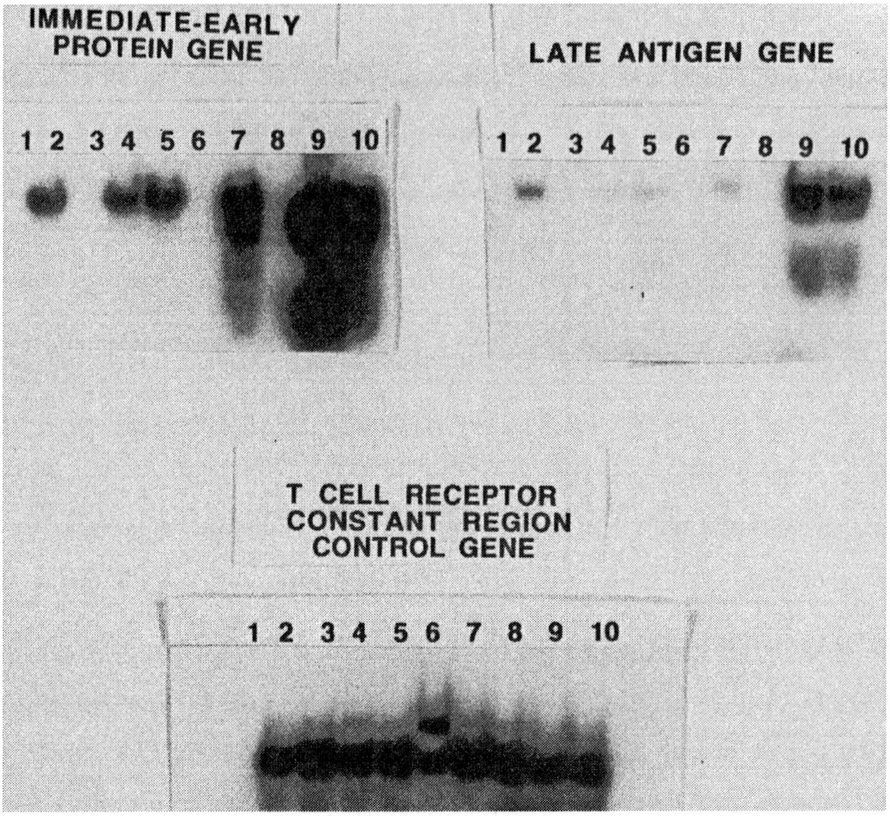

Figure 1. Autoradiogram showing the liquid hybridization/gel retardation analysis of amplified cytomegalovirus DNA in the leukocytes of renal transplant recipients. Upper left panel represents amplification of the immediate early gene. Upper right panel represents amplification of the late antigen gene. Bottom panel shows amplification using the TcR/βc primers as evidence that the DNA extract from patient specimens was adequate to support amplification. Lane 1, reagent control (no DNA), Lanes 2-7, patient specimens, Lane 8, negative DNA control, Lanes 9 and 10, positive CMV DNA controls.

left panel and the late antigen gene in the top right panel. In addition, all six patient specimens were amplified by the TcR/βc control primers as proof that the DNA extracted from each specimen was amplifiable and did not contain inhibitors of the amplification reaction.

Table II illustrates the concordance of the two assays in detecting CMV viremia. The PCR assay detected several positive specimens that were not detected by antigenemia. Suggesting that the PCR may represent an increase in sensitivity over the antigenemia assay.

Table III contains all the data collected from the renal transplant patients and reveals that the PCR is also more consistently positive in consecutive specimens from a single patient, as is seen with patient #3, 7 and 9. In specimens from patients #9, 12, and 13 the PCR was positive 16, 9, and 20 days earlier respectively than the antigenemia assay, but all three patients were subsequently positive by both assays. Patients #12 and 13 were positive by both assays for a period of two months. Patient #9 was consistently positive by PCR over a four month period, while the antigenemia assay detected CMV viremia sporadically over a two month period.

Patient #7 was detected as viremic by both assays on the same specimen, but the PCR continued to be positive for 20 days after the antigenemia assay reverted to negative. Thus it appears that the PCR assay is slightly more sensitive than the antigenemia assay.

Three patients, # 4, 10 and 17 were positive by PCR but not detected as viremic by the antigenemia assay. In each case the PCR was positive in either the only specimen available from this patient (#17), or in the last specimen collected from that patient (#4 & 10). Patient #4 had experienced a CMV viremic period, as diagnosed by antigenemia, for three months beginning four months before these specimens were drawn and was shedding virus in the urine at the time the PCR was positive and over a previous two

TABLE II

CLINICAL DETECTION OF CMV IN PERIPHERAL BLOOD LEUKOCYTES OF RENAL TRANSPLANT RECIPIENTS

Patient Number	Antigenemia Results[1]	PCR Results[1]	Months Post Transplant	Treatment[2]
1	4/6	6/6	5+	None
2	2/3	2/3	7	None
3	5/8	8/8	5	Yes -3
4	0/5	2/5	5	Yes -3-5
5	12/13	13/13	4	Yes -2
6	0/7	0/7	3	None
7	4/10	7/10	3	Yes -2
8	0/3	0/3	<1	None
9	3/8	7/8	1	None
10	0/2	1/2	1	None
11	0/5	0/5	0[4]	None
12	8/11	10/11	0	Yes - same
13	11/14	13/14	0	Yes +1
14	4/5	4/5	0	None
15	0/3	0/3	3	None
16	0/1	0/1	3	None
17	0/1	1/1	6	None
18	0/1	0/1	NA[3]	NA
19	1/1	1/1	NA	NA
20	0/1	0/1	NA	NA
21	0/1	0/1	NA	NA
22	0/1	0/1	NA	NA
23	0/1	0/1	NA	NA
24	0/1	0/1	NA	NA
25	1/1	1/1	NA	NA

[1] Positive / Number Tested
[2] Cytogam plus Gancyiclovir
[3] Not Available
[4] Simultaneous

week period. Patient #10 was reassayed by antigenemia and was declared +/- on the same date that the PCR was positive and had been positive by antigenemia three months earlier. No further information or additional specimens were available for patient # 17.

The discrepancies between the antigenemia assay and the PCR do not appear to be related to therapy. Patients # 3, 4, 5, 7, 12, and 13 all received cytogam and ganciclovir therapy. Treatment was initiated up to two months prior to the specimens tested here (#3, 4, 5, and 7). Patients #12 and 13 were treated with cytogam and ganciclovir during the two month period that both assays were repeatedly positive for CMV.

It is important to note that eleven patients were negative for CMV by both assays, of these #6, 8, 11, and 15 were assayed repeatedly, as many as seven times with complete agreement between the two assay systems.

The PCR protocol employed in this study was very sensitive, quantitative data indicated that as little as 0.6 $TCID_{50}$ of CMV DNA could be detected (20). However these assays required 24 - 48 hours to complete, and the routine use of radioactivity in a clinical laboratory setting poses some hazard to the employees as well as to the environment. The utilization of a fluorescent detection system, would avoid the hazards of radioactivity, if comparable sensitivity of detection could be achieved.

A protocol has been developed, using the fluorescent intercalating dye SYBR Green I, and a laser scanner (20). The FluorImager employs an argon laser to detect fluorescence in multiple formats, including acrylamide gels,

TABLE III

Patient # Date[2]	AGN[1]	PCR
1. (6)		
12/27	+/-	+
1/20	+	+
3/8	+	+
3/14	+	+
3/28	+	+
4/11	+/-	+
2. (3)		
2/22	--	--
3/8	+	+
3/15	+	+
3. (8)		
12/13	+	+
12/27	+	+
1/11	--	+
1/25	+	+
2/9	+	+
3/1	+/--	+
3/23	+	+
4/13	+/--	+
4. (5)		
12/14	--	--
12/27	--	--
1/18	--	--
2/23	--	+
3/1	--	+
5. (13)		
12/27	+	+
1/10	+	+
1/20	+	+
1/24	+	+
1/31	+	+
2/7	+	+
2/28	+	+
3/7	+	+
3/14	+	+
3/28	+	+
4/5	+	+
4/11	+	+
4/25	+/--	+

Patient # Date	AGN	PCR
6. (7)		
12/13	--	--
12/27	--	--
1/10	--	--
1/24	--	--
2/7	--	--
2/23	--	--
3/21	--	--
7. (10)		
12/13	--	--
12/27	+	+
1/3	+/--	+
1/12	+	+
1/26	+	+
2/9	--	--
3/8	+	+
3/14	--	+
3/28	--	+
4/27	--	--
8. (3)		
3/29	--	--
4/11	--	--
4/13	--	--
9. (8)		
12/13	--	+
12/29	+	+
1/5	--	+
1/11	+	+
1/20	+	+
2/15	+	--
3/2	--	+
4/6	--	+
10. (2)		
12/27	--	--
2/22	--	+
11. (5)		
1/25	--	--
2/7	--	--
3/23	--	--
4/5	--	--
4/26	--	--

TABLE III Continued

12. (11)		
2/8	--	--
3/8	--	+
3/17	+	+
3/23	+	+
3/28	+	+
3/30	+	+
4/4	+	+
4/6	+	+
4/13	+	+
4/25	+	+
4/27	+/--	+
13. (14)		
2/7	--	--
2/23	--	+
2/28	--	+
3/14	+	+
3/16	+	+
3/23	+	+
3/28	+	+
3/30	+	+
4/4	+	+
4/6	+	+
4/11	+	+
4/13	+	+
4/15	+	+
4/25	+	+
14. (5)		
2/8	--	--
3/21	+	+
3/28	+	+
4/4	+	+
4/20	+	+
15. (3)		
1/5	--	--
1/26	--	--
3/9	--	--
16. (1)		
1/24	--	--
17. (1)		
1/24	--	+
18. (1)		
1/7	--	--
19. (1)		
1/26	--	+
20. (1)		
2/14	--	--
21. (1)		
2/14	--	--
22. (1)		
2/15	--	--
23. (1)		
3/21	--	--
24. (1)		
4/26	--	--
25. (1)		
4/26	+	+

[1] AGN - Antigenemia

1) [2] Date specimen collected

agarose gels, and 96 well plates. Figure 2 illustrates the protocol now in use for the detection of amplified DNA. Specimens from many sources, including blood, spinal fluid, nasopharyngeal aspirates and swabs, amniotic fluid and tissue biopsy specimens are solubilized in the same buffer systems described for the radioactive assays. Extracted DNA is amplified using gene specific primers as before, and the amplified DNA is mixed with SYBR Green I and

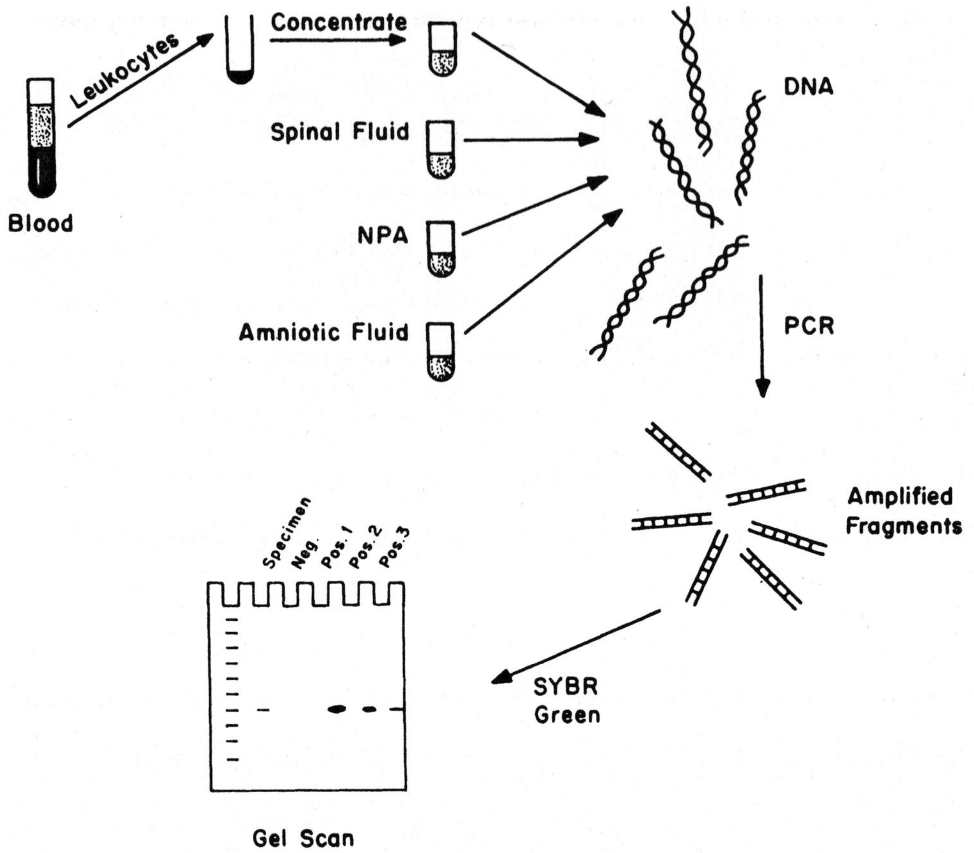

Figure 2. Schematic diagram of the fluorescent intercalating dye method of detecting amplified DNA from a variety of specimen sources.

electrophoresed on an acrylamide gel. The gel, still between two glass plates is inserted directly into the FluorImager. Scanning a 16 x 20 cm gel requires less than three minutes. The image is projected on a computer screen and can be enhanced and analyzed statistically. This assay requires approximately 6 hours to complete, from time of receipt of the specimen to reporting of results (20). Quantitative analysis of CMV DNA using these methods indicated that 0.06 $TCID_{50}$ of CMV DNA could be detected, thus the method is equally as sensitive as the radioactive detection system, and perhaps somewhat more sensitive.

Figure 3 shows an image of a fluorescent stained gel including amplified DNA from peripheral blood leukocytes of three bone marrow transplant recipients. Patients A and C were negative for cytomegalovirus, Patient B was positive for cytomegalovirus, showing a band in both duplicates that migrates at the same rate as the positive control CMV DNA in the same gel. Reagent control (no DNA) and negative control DNA amplified in the same reaction system did not show bands migrating with the positive controls. Patients DNA amplified with the primers for TcR/βc show bands migrating at the same rate as DNA amplified from the negative control DNA which is human leukocyte DNA. This is evidence that the integrity of the DNA extracts from patient specimens was sufficient to support uninhibited amplification.

DISCUSSION

The detection of CMV in peripheral blood leukocytes by PCR linked with liquid hybridization and gel retardation provides a very sensitive assay for the presence of this virus. Quantitative measurement of the infectious virus detected by this method revealed that less than one $TCID_{50}$ (0.6 $TCID_{50}$) was detectable (20). Application of this method to the clinical specimens obtained from renal transplant patients in comparison with the antigenemia assay showed remarkable concordance between the two assays. The polymerase chain reaction assay sometimes detected an infection

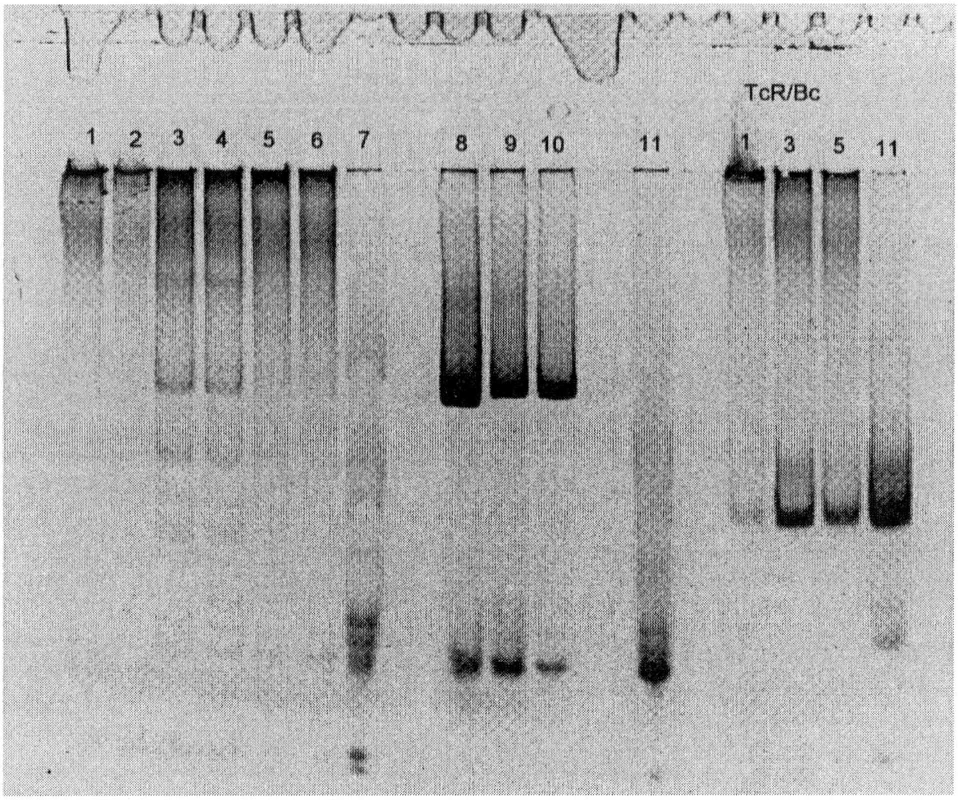

Figure 3. SYBR Green I stained acrylamide gel of DNA extracted from peripheral blood leukocytes of bone marrow transplant recipients and amplified for cytomegalovirus with controls. Lanes 1&2, Patient A; Lanes 3&4, Patient B; Lanes 5&6, Patient C; Lane 7, reagent control (no DNA); Lane 8, CMV DNA at 14 fg; Lane 9, CMV DNA at 7 fg; Lane 10, CMV DNA at 3.5 fg; Lane 11, negative DNA control.

(defined by viremia) as much as two weeks earlier than the antigenemia assay. A recent study has indicated that two consecutive positive PCR assays, when conducted weekly to monitor bone marrow transplant recipients could be used to initiate antiviral treatment to prevent CMV disease (21). Also, the PCR sometimes remained positive for longer time intervals after an infectious episode than did the antigenemia assay. These results suggest that the PCR is slightly more sensitive than the antigenemia assay for the detection of CMV in peripheral blood leukocytes.

The data do not suggest that the PCR is detecting false positive infections, since the few cases where antigenemia was negative and PCR was positive were subsequently found to be either +/- on reassay by antigenemia, or evidence of virus shedding at the time of sampling suggested an active infection.

The PCR assay provides an advantage over the antigenemia assay, regardless of sensitivity, because it is applicable to many specimen types, not just peripheral blood leukocytes. These data as well as the results of other investigators (15-18) indicate that the PCR assay has significant advantage over the antigenemia assay.

One problem in application of this methodology to the clinical laboratory setting is that use of radioactivity under conditions were multiple samples are being analyzed on a regular bases poses some hazards to the laboratory personnel and to the environment. Also, the decay of radioactive probes, requires frequent relabeling and makes comparison of gels run at different times difficult. The development of an assay system using fluorescent detection of amplified DNA solves these problems. Studies comparing the sensitivity of detection of viral DNA using ^{32}P hybridization and gel retardation compared to fluorescent staining and gel imaging have indicated that the fluorescent method is as sensitive, perhaps slightly more sensitive than the radioactive detection methods (20). In addition the fluorescent assays can be completed more quickly, within six hours of receipt of the specimen.

REFERENCES

1. Ackerman, J.R., Lefor, W.M., Weinstein, L, et. al. Transplant. Proc. 20(S1):468-71. (1988)
2. Peterson, P.K., Balfour, H.H., Masker, S.C., et. al. Medicine 59:283-300. (1980)
3. Rubin, R.H. Clinical Approach to Infections in the Compromised Host, 2nd ed, New York: Plenum, p. 557. (1988)
4. van der Bij, W., Schirm, J., Torensman, R., et. al. J. Clin. Microbiol. 26:2531-35. (1988)
5. van der Bij, W., Torensman, R., van Son, J., et. al. J. Med. Virol. 25:1790-88. (1988)
6. Bamborschke, S., Wullen, T., Huber, M., Neveling, M., Baldamus, C.A., Korn, K. and John, G. 1992. J. Neurol. 239:205-8. (1992)
7. Drouet, E., Michelson, S., Denoyel, G., Colimon, R. J. Virol. Methods 45:259-276. (1993)
8. Einsele, H., Steidle, M., Vallbracht, A., Saal, J.G., Ehninger, G., Muller, C.A. Blood 77:1104-10. (1991)
9. Einsele, H. Ehninger, G., Steidle, M. et.al. Lancet 338:1170-72. (1991)
10. Lee, W.T., Antoszewska, H. Powell, K.F., et. al. 1992. Aust. N.Z. J. Med. 22:249-55.(1992)
11. Miller, M.J., Bovey, S., Pado, K., Bruckner, D.A., and Wagar, E.A. J. Clin. Microbiol. 32:5-10. (1994)
12. Rowley, A.H., Wolisnky, S.M., Sambol, S.P., Barkholt, L., Ehrnst, A., Andersson, J.P. Transplant. 51:1028-33. (1991)
13. Spector, S.A., Srill, R., Wolf, D., Dankner, W.M. J. Clin. Microbiol. 30:2359-65. (1992)
14. Stanier, P., Kitchen, A.D., Taylor, D.L., Tyms, A.S. Molec. Cell Probes 8:51-58. (1992)
15. Gerna, G., Zipeto, D., Parea, M. et.al. J. Infect. Dis. 164:488-498. (1991)
16. van Dorp, W.T., Vlieger, A., Jew, N.M., et. al. Transplant. 54:661-64. (1992)
17. Vlieger, A.M., Boland, G.J., Jewa, N.M. et. al. 1992. Bone Marrow Transplant 9:247-53. (1992)

18. Zipeto, D., Revello, M.G., Silini, E., et. al. J. Clin. Microbiol. 30:527-30.
19. Gerna, G., Revello, M.G., Percivalle, E., Morini, F. 1992. J. Clin. Micro. 30:1232-37. (1992)
20. Evans, M.J., Edwards-Spring, Y., Povinelli, D., Wendt, A., Harvey, R., Rittenhous-Diakun, K. and Riepenhoff-Talty, M. Clin. Diag. Virol. 6:163-74. (1996)
21. Evans, M.J., Evans, R.T. and Greenberg, S.J. PCR-Based Diagnostics in Infectious Diseases, Blackwell Scientific Publications, Boston, Chap 32. pp. 609-19. (1994)
22. Yuen, K.-Y., Lo, S.K.-F., Chiu, E.K.-W., Wong, S.S.-Y. J. Clin. Microbiol. 33:2530-34. (1995)

HTLV-ASSOCIATED DISEASES: HUMAN RETROVIRAL INFECTION AND CUTANEOUS T-CELL LYMPHOMAS

K. Fujihara, B. Goldman, A.R. Oseroff, N. Glenister, E.S. Jaffe,
E. Bisaccia, S. Pincus, S.J. Greenberg
Departments of Neurology and Dermatology, Roswell Park Cancer Institute
and State University of New York at Buffalo, Buffalo, New York 14263,
National Cancer Institute, National Institutes of Health, Bethesda, Maryland 20892-4470,
and the Department of Dermatology, Morristown Memorial Hospital,
Morristown, New Jersey 07960

ABSTRACT

An array of neurologic, oncologic, and autoimmune disorders are associated with infection with the human pathogenic retroviruses human T-cell leukemia virus types I and II (HTLV-I, II), as well as the human immunodeficiency viruses (HIV). The cutaneous T-cell lymphomas, mycosis fungoides (MF) and its hematogenous variant Sezary Syndrome (SS), share similar clinical and pathological features to HTLV-I-associated adult T-cell leukemia (ATL) and speculation of a retroviral link to MF and SS, especially in areas non-endemic for ATL, has lead to an intensified search for HTLV- and HIV-like agents in these diseases. To further explore a potential role for human retroviruses in MF and SS, skin biopsy-derived or peripheral blood mononuclear cell-derived DNA from 17 patients (MF, n=7; erythrodermic MF (EMF), n=5; SS, n=5) from the North Eastern United States were screened using gene amplification by PCR and a liquid hybridization detection assay. Previously published primers and probes for HTLV-I (LTR, *gag*, *pol*, *env*, and *pX*), and our own primers and probes for HTLV-I (*gag*, *pcl*, and env), HTLV-II (*pol* and *env*) and HIV-I (*gag* and *pol*) were employed. Serum antibodies to HTLV-I were negative in all but one EMF patient. The single HTLV-I seropositive patient carrying a diagnosis of EMF generated positive amplified signals for all of the eight HTLV-I regions tested. Ultimately, this individual evolved to exhibit clinical manifestations indistinguishable from ATL. The other 16 patients were negative for all 12 HTLV and HIV retroviral regions. Our findings suggest that none of the known prototypic human retroviruses are associated with seronegative MF and SS. The uniformly positive results for HTLV-I in the seropositive patient suggests that this patient initially presented with a smoldering form of ATL and illustrates the difficulty that sometimes may be encountered in the differential diagnosis of MF, SS, and ATL based solely on clinical and histopathological criteria.

Address correspondence to: *Steven J. Greenberg, MD, Laboratory of Neuroimmunology & Neurovirology, Department of Neurology, Roswell Park Cancer Institute, Elm and Carlton Streets, Buffalo, New York 14263. Phone: (716) 845-8600. Fax: (716) 845-8656.*

Key Words: Retrovirus; HTLV-I; HTLV-II; HIV; Lymphoma, T-cell, Cutaneous; PCR

INTRODUCTION

One of the more compelling applications of gene amplification by polymerase chain reaction (PCR) has been its usefulness in screening various patient-derived specimens for an array of infectious agents. In addition to PCR-based molecular diagnostics for viral infections, several large scale epidemiological studies have been conducted in an attempt to establish a correlation between prototypic virus based on PCR positivity and certain disease entities, including multiple sclerosis (1) and inflammatory myopathies (2). Because they share some clinical and pathological features (3-5), there has been considerable speculation about a potential HTLV-I-related etiology in MF and SS. The association of other retroviruses, HTLV-II and HIV, with MF and SS, also has been suggested (6,7). However, a retroviral role in the pathogenesis of MF and SS remains controversial (8-12). The presence of "deleted" HTLV-I proviral sequences in an HTLV-I seronegative American patient with SS and in seronegative Swedish patients, four with MF and one with SS, has been reported (13). However, these observations have raised considerable concerns about the manner in which the HTLV-I-like sequences were detected in MF/SS. In the American patient, "deleted" HTLV-I was demonstrated by Southern hybridization against DNA extracted from a *non-T*, Epstein-Barr virus-transformed B-cell line derived from the patient's blood. The cells underwent considerable manipulation and were first expanded with interleukin-2 and phytohemagglutinin and subsequently co-cultivated with cord blood cells. Restriction fragment length polymorphism and sequence analyses revealed multiple endonuclease restriction site alterations and a major 5.5 Kb deletion within the prototype HTLV-I proviral sequence. Thus, the retroviral sequences detected represented only limited regions of the HTLV-I genome, and it remains unclear to what degree the DNA from the B-cell line reflected the characteristics of the malignant T-cells in blood or in cutaneous lesions. Regrettably, there was no attempt to amplify blood lymphocyte DNA *ex vivo* from the patient.

In the Swedish patients, the technique of gene amplification by PCR was used to screen MF/SS DNA from cutaneous lesions and blood for five separate regions of the prototype HTLV-I provirus. All five Swedish MF/SS patients tested positive for HTLV-I-related sequences, but the patterns of amplification (LTR, *gag*, *pol*, *env*, or *pX*) varied widely among patients and between paired skin and blood samples. Negative results of Southern hybridization analyses of two DNA samples were thought due to low proviral copy number. Unfortunately, no attempt was made to corroborate this conclusion by quantitative PCR nor was there an attempt to confirm whether the amplicons were truly derived from integrated retroviral sequences. In our experience, inadvertent amplification of carry-over amplicon is easily identified by revising the PCR to include one or a set of primers that is externally juxtaposed to the original primer set. While a negative signal does not exclude an integrated, truncated proviral sequence, a positive signal strongly corroborates the original PCR findings. This test was not performed. Thus, although intriguing, this prior study may be compatible with artifacts introduced by the methodology, as experienced in other studies.

To assess the frequency by which one might expect to find prototypic retroviral sequences among MF/SS patients, we performed an exhaustive screening of DNA from 17 patients from the Northeastern United States for the presence of HTLV-I, HTLV-II, and HIV sequences by PCR.

MATERIALS AND METHODS

Subjects. Three groups of patients comprising MF (n=7), erythrodermic mycosis fungoides (EMF) (n=5), and SS (n=5) were studied. One patient (EMF-4) was found to be seropositive for antibodies to HTLV and the other 16 patients were seronegative. Diagnoses were confirmed by clinical and histopathological examinations. HTLV serology was tested by a commercially available ELISA kit. The HTLV-I-seropositive patient (EMF-4) was a 35-year-old black female born and bred in the United States without apparent risk factors for acquiring HTLV. Serum γ-globulin (29.4%, normal range [11.6 - 22.5%]) and lactate dehydrogenase (666 U/ml, [283-555 U/ml]) were slightly elevated. Serum soluble interleukin-2 receptors (2676 U/ml, [240-600 U/ml]) were increased. Her skin lesions responded well to photopheresis therapy. PCR analyses of blood were performed in 13 patients, analyses of skin lesions in 8 patients, and analyses of both blood and skin in 4 patients.

DNA and Oligonucleotide Preparation. Peripheral mononuclear cells (PBMNC) were isolated from whole blood by density gradient centrifugation and high molecular weight DNA was extracted from PBMNC by the sodium dodecyl sulfate/proteinase K method, purified by pheonol/chloroform extraction, precipitated in absolute ethanol, and resuspended in 10:1 TE (10 mmol/L Tris-HCl and 1 mmol/L EDTA) buffer as previously described (14,15). Biopsied skin lesions were snap frozen in liquid nitrogen and specimens were first crushed with mortar and pestle and then subjected to DNA extraction, as above, except that DNA extraction from skin lesions was performed using a non-ionic detergent and proteinase K digestion (16).

Oligonucleotide primers and probes were synthesized at the Biopolymer Core Facility, Roswell Park Cancer Institute (CA16056-1) by the phosphoramidite method on an automated 391 DNA synthesizer (Applied Biosytems, Foster City, CA) and 5' trityl-retained oligonucleotides were purified by chromatography using Nensorb Prep™ cartridges (Dupont, Boston, MA) and then concentrated by lyophilization under vacuum.

A total of 12 sets of primers and probes for HTLV-I, HTLV-II, and HIV were employed, targeting eight, two, and two separate retroviral regions, respectively. Previously reported primers and probes for HTLV-I (LTR, *gag, pol, env,* and *pX* regions) as reported by Hall, et al. (13) were used, as well as three of our own previously established primer stets (17). The following retroviral primers were developed and validated in our laboratory: HTLV-I/*gag* (1374-1407 and 1939-1906), HTLV-I/*pol* (4179-4209 and 4628-4600), HTLV-I/*env* (5262-5296 and 5816-5784), HTLV-II/*pol* (4735-4756 and 4921-4899), HTLV-II/*env* (5799-5818 and 6126-6107), HIV/*gag* (1544-1571 and 1659-1632) and HIV/*pol* (4287-4326 and 4426-4390). Corresponding internal probes were as follows: HTLV-I/*gag* (1661-1700); HTLV-I/*pol* (4552-4591), HTLV-I/*env* (5612-5650), HTLV-II/*pol* (HTLV-II/*pol* (4880-4898), HTLV-II/*env* (5841-5880), HIV/*gag* (1588-1628), and HIV/*pol* (4336-4376). Position numbers for our primers and probes for HTLV-I, HTLV-II, and HIV are based on Genbank's sequences for loci HL1PROP, HIVV2CG, and HIVNL43, respectively.

Gene Amplification and Amplicon Detection by Liquid Hybridization. All PCRs were carried out against 1 μg of DNA template in 100 μl adjusted to final concentrations of

10 μM KCl, 1.5 mM $MgCl_2$, 0.001% gelatin, 200 μM deoxynucleotide triphosphates, and 200 nM each primer to which 0.8 unit Taq polymerase (Perkin-Elmer Cetus, Norwalk, CT) was added. Temperature cylcing was carried out in a DNA Thermal Cycler (Perkin-Elmer Cetus) as follows: An initial denaturation for 7 min at 94°C and then 1 min at 94°C, primer annealing at 55°C for one minute, and 2 min at 72°C for primer extension for 30 cycles.

Thirty microliters of each PCR product was added to a hybridization mix containing ^{32}P-end labeled probe ($2.5x10^5$ cpm), and 5 μl 1.5M NaCl to a final volume of 50 μl, heat denatured at 98°C for five minutes, and then allowed to anneal at 55°C for 60 minutes. The samples were run on 5% polyacrylamide gel (200 V, 1 hour), and the gels were exposed to X-OMAT AR films overnight (15-20 hours) at room temperature.

Cloning and Sequencing. The amplified product was separated from minor contaminating bands and excess PCR primers by subjecting the amplicon to filtration centrifugation using ultra-free-MC filter units, 30,000 molecular weight cut off (Millipore, Bedford, MA). The purified, amplified sequence was directly ligated into pCR II™ plasmid (Invitrogen, San Diego, CA). Subsequent to ligation and bacterial transformation into INV a F^1 cells (Invitrogen, San Diego, CA), the transformants were selectively expanded in Kanamycin containing agar and productive clones were selected by extinction of the β-galactosidase marker. The insert-containing plasmids were extracted from bacterial transformants by the plasmid miniprep technique, and the HTLV-containing segments were direclty sequenced from the purified plasmid by the dideoxy chain termination method (18,19). Sequencing primers span either the T7 promoter or the SP6 promoter plasmid regions. The other reagents used for the DNA sequencing reactions were supplied with the USB Sequencing Kit (United States Biochemical Corp., Cleveland, OH). At least three separate clones were fully sequenced to obtain a consensus genetic sequence.

RESULTS

None of the 16 seronegative patients generated positive signals for the prototype retroviral sequences tested, regardless of disease type or whether skin or blood DNA was analyzed (Fig. 1, A~N). The HTLV-I-seropositive patient (EMF-4) was positive for 7

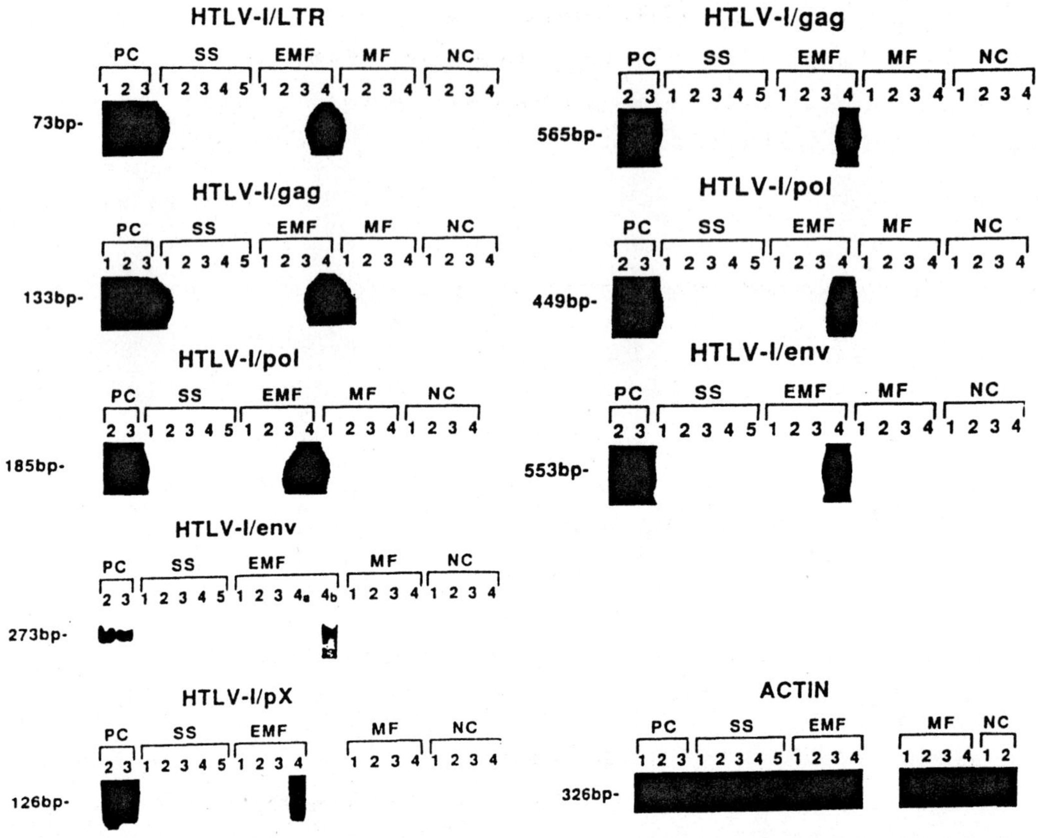

FIGURE 1. Screening for HTLV-I in Cutaneous T-Cell Lymphoma. Autoradiographs of gel retardation detection assays following liquid hybridizations were performed on patient DNA derived from peripheral blood mononuclear cells. *Left Column:* The previously reported primers and probes (13) were used, HTLV-I/LTR, *gag*, *pol*, *env*, and *pX*. *Right Column:* Our own primers and probes (17) were used, HTLV-I/*gag*, *pol*, and *env*. SS-1-5, EMF-1-4, and MF-1-4 denote each patient tested. Positive controls included HUT102 (PC-1), MT-1 (PC-2), and an ATL patient (PC-3). Negative controls included the T-cell line MOLT-4 (NC-1), a seronegative volunteer (NC-2), salmon sperm DNA (NC-3), and PCR reaction mix without DNA (NC-4). Except for EMF-4, none of the patient blood specimens tested positive for retroviral sequences. In addition, samples of skin biopsy-derived DNA (SS-2,-3, EMF-5, and MF 2, 3, 5, 6, and 7) were subjected to PCR-based analysis for HTLV-I (*pol* and *gag*), HTLV-II (*env*) and HIV *(pol)*. All biopsied skin specimens were negative (data not shown). Screening for one HTLV-I/*env* region in patient EMF-4 produced variable results in which one experiment was negative (EMF-4a), but another positive (EMF-4b). However, in EMF-4b, the PCR product was of expected size (273 bp) on 1% ethidium bromide-stained agarose gel, suggesting variation in the region corresponding to the probe sequence.

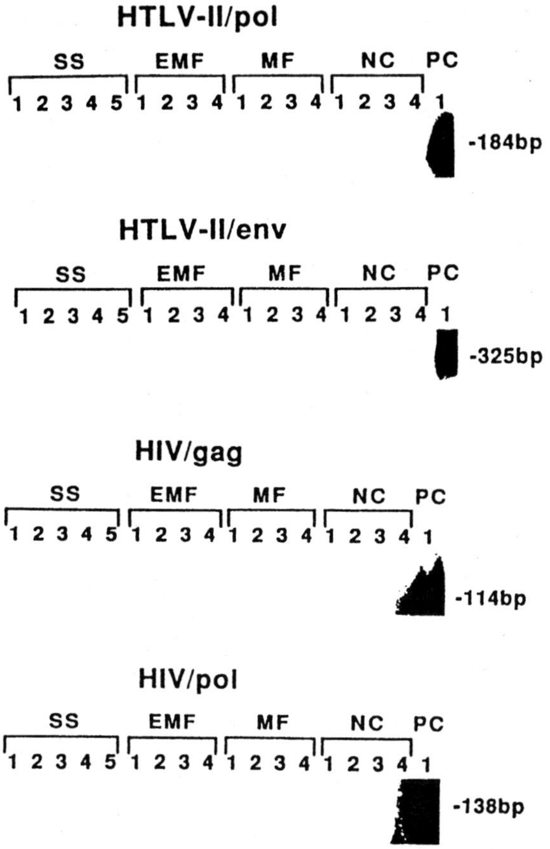

FIGURE 2. Screening for HTLV-II and HIV in Cutaneous T-Cell Lymphoma. Blood-derived DNA on all patients and skin-derived DNA (SS-2 and -3, EMF-5, and MF-2,-3,-5, -6, and -7) were subjected to gene amplification and the amplicons analyzed by liquid hybridization and gel retardation electrophoresis. Representative samples are displayed. All reactions failed to detect the presence of HTLV-II or HIV retroviral sequences. Positive control (PC) for HTLV-I consisted of DNA derived from the T-cell line MO-T and positive control (PC) for HIV consisted of DNA derived from the T-cell line 8E5/LAV.

of 8 HTLV-I regions tested; an HTLV-I/env targeted region used previously (13) generated an inconsistent signal in our detection system. Since the amplified HTLV-I/*env* region of the HTLV-I seropositive patient (EMF-4) was of expected size (273 bp) on 1% ethidium bromide-stained agarose gel (Fig. 1, 0), point mutations in this region were suspected. Six point mutations were identified in the 274 bp region: (5468) C→T, (5472) G→A, (5610) A→G, (5613 C→A, (5637) T→G, and (5667) A→G. Two point mutations

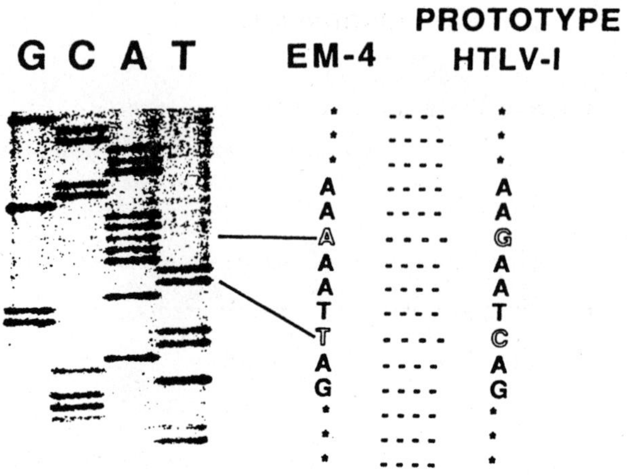

FIGURE 3. Sequence Analysis of EM-4 Patient HTLV-I/env Amplified Sequence. Sequence analysis revealed 6 point mutations in the env target region, including two-point mutations: (5468) C→T and (5472) G→A, in the probe-hybridizing sequence (5455-5483).

at 5468 and 5613, caused changes of deduced amino acid sequences, Threonine→Isoleucine, and Histidine→Glutamine, respectively.

The skin biopsy of the HTLV-I seropositive patient (EMF-4) was specifically reviewed. Sections demonstrated a band-like infiltrate in the superficial dermis. In addition, there was infiltration of the overlying epidermis by individual cells, although well formed Pautrier's microabscesses were not present. Smaller lymphocytes showed atypia and some appeared cerebriform. The histologic fidings of atypical dermal infiltration with epidermotropism were consistent with the diagnosis of mycosis fungoides.

We attempted to analyze the nature of proviral integration into host genome of the seropositive, PCR-positive patient by Southern hybridization analysis. However, no proviral restriction length fragment band was identified from blots obtained from BamHI and HindIII or EcoRI digests (data not shown). Based on data in a quantitative PCR, it was estimated that about 0.1% of EMF-4 PBMC were infected with 1 copy of HTLV-I provirus. This low frequency of circulating HTLV-I-infected cells is below the level of sensitivity of Southern hybridization analysis.

DISCUSSION

Thus, by exhaustive molecular diagnostics employing gene amplifications, we failed to detect HTLV-I proviral sequences in the skin or blood of the 16 seronegative patients. Insensitivity of our assay system is not a likely explanation, since we were able to detect approximately a single copy of HTLV-I provirus per million cells by quantitative PCR. Alternatively, a small subset of ATL patients are seronegative, but positive for HTLV-I by PCR (20,21). However, that all five of the Swedish MF/SS patients previously studied would represent this ATL subgroup would be remarkable.

There are multiple reasons for PCR pattern variability in population screening studies in which a panel of target sequences are studied and need not imply major sequence deletions in target regions. Simple point mutations corresponding to the most 3' primer region can negate efficient PCR, as has been observed in PCR amplification of an HTLV-I/*env* region (22). Limited polymorphisms corresponding to bases comprising the internal oligonucleotide probe may also account for a false negative result, as shown in patient EMF-4a in this study (Fig. 1). These potential pitfalls are largely overcome by choosing primers and probes that flank and recognize highly conserved regions within the HTLV-I genome. Our own HTLV-I primers and probes to *gag*, *pol*, and *env* have been tested extensively against patients with ATL, HTLV-I-associated myelopathy/tropical spastic paraparesis, and HTLV-I seropositive carriers, and are uniformly highly effective in detecting prototype HTLV-I by PCR (23). Using our three primer sets in addition to the identical five primer sets previously reported (13) (a total of eight separate regions targeting the HTLV-I genome), we failed to detect the presence of HTLV-I-like sequences in seronegative MF/SS. Two regions specific for HTLV-II and two regions specific for HIV were also tested and were negative in all patients. These negative results do not exclude the possibility of a retroviral etiology in MF and SS, but strongly suggest that none of the known prototypic human pathogenic retroviruses are commonly associated with MF and SS.

The patient EMF-4, who carried a presumptive diagnosis of erythrodermic mycosis fungoides, was found to be seropositive for antibodies to HTLV and was positive by PCR for HTLV-I. We conclude that this patient may be exhibiting a smoldering form of ATL (24-27). This case is an example illustrating the difficulties that are sometimes encountered in the differential diagnosis between MF/SS and ATL made solely by clinical and histological criteria (24). This situation argues strongly for retroviral screening via serology and PCR analysis in patients suspected of having MF/SS.

ACKNOWLEDGEMENTS

This work was supported in part by a grant from the Endowment for the Neurosciences.

The authors wish to thank S. Armus, F. Dispattro, and H. Stoll for assistance in providing some of the clinical specimens. T.-L. Du, S. Nagafuchi, M. Rickert, and P. Overturf are recognized for providing technical assistance. Finally, the authors are indebted to G. Lombardo for providing excellent secretarial support in the preparation of this manuscript.

REFERENCES

1. G.D. Ehrlich, J.B. Glaser, V. Bryz-Gornia, J. Maese, T.A. Waldmann, B.J. Poiesz, and S.J. Greenberg, Neurology, 41, 335-343 (1991).

2. R.L. Leff, L.A. Love, F.W. Miller, and S.J. Greenberg, E.A. Klein, M.C. Dalakas, and P.H. Plotz, Lancet 339, 1192-1195 (1992).

3. P.A. Bunn, Jr., S.J. Hoffman, D. Norris, L.E. Golitz, J.L. Aeling, Annals of Internal Medicine, 309, 257 (1993).

4. L. Ratner and B.J. Poiesz., Medicine, 67, 401-422, (1988).

5. A. Chadburn, E. Athan, R. Wieczorek, and D.M. Knowles, Blood, 77, 2419-2430 (1991).

6. D. Zucker-Franklin, W.C. Hooper, and B.L. Evatt, Blood, 80, 1537-1545 (1992).

7. G.A. Crane, D. Variakojis, S.T. Rosen, A.M. Sands, and H.H. Roenigk, Jr., Arch. Dermatol., 127, 989-994 (1991).

8. V. Manzari, A. Gismondi, G. Barillari, S. Morrone, A. Modesti, L. Albonici, L. De Marchis, V. Fazio, A. Gradilone, and M. Zani, Science, 238, 1581-1583 (1987).

9. C. Capésius, F. Saal, E. Maero, A Bazarbachi, J. Lasneret, L. Laroche, A Gessain, F.Hojman, and J. Peries, Leukemia, 5,416-419 (1991).

10. B. Hjelle, R. Mills, S. Swenson, G. Mertz, C. Key, and S. Allen. J. Infect. Dis., 163, 435-440 (1991).

11. D. Zucker-Franklin, E.E. Coutavas, M.G. Rush, and D.C. Zouzias., Proc. Natl. Acad. Sci. (USA), 88, 7630-7634 (1991).

12. G. Lisby, M.S. Reitz, Jr., and G.L. Vejlsgaard., J. Invest. Dermatol., 98, 417-420 (1992).

13. W.W. Hall, C.R. Liu, O. Schneewind, H. Takahoshi, M.H. Kaplan, G. Röupe, A. Vahlne., Science, 253, 317-320 (1991).

14. A. Maluish and D.M. Strong, Manual of Clinical Laboratory Immunology, N.R. Rose, H. Friedman, and J.L. Fahey, eds., American Society for Microbiology, Washington, DC, (1986) pp. 277-278.

15. J. Sambrook, E.F. Fritsch, and T. Maniatis. Molecular Cloning: A Laboratory Manual, Cold Spring Harbor Laboratory Press, Cold Spring Harbor, New York, (1989) pp. 9.16-9.19.

16. R. Higuchi. PCR Technology: Principles and Applications for DNA Amplification, H. Erlich, ed., Stockton Press, New York, (1989) pp. 31-38.

17. G.D. Ehrlich and S.J. Greenberg, PCR-Based Diagnostics in Infectious Disease, Blackwell Scientific, Cambridge, Massachusetts, (1994) pp. 415-446.

18. J. Sambrook, E.F. Fritsch, and T. Maniatis. A Laboratory Manual, Cold Spring Harbor Laboratory Press, Cold Spring Harbor, New York, (1989) p. 1.25.

19. F. Sanger, S. Miklen, and A.R. Coulson, Proc. Natl. Acad. Sci. (U.S.A.), 74, 5463-5467 (1977).

20. D.B. Duggan, G.D. Ehrlich, F.P. Davey, S. Kwok, J. Sninsky, J. Goldberg, L. Baltrucki, and B.J. Poiesz, Blood 71, 1027-1032 (1986).

21. S. Kwok, G. Ehrlich, B. Poiesz, R. Kaliksh, J.J. Sninsky, Blood 72, 1117-1123 (1988).

22. G.D. Ehrlich, J. Andrews, M.P. Sherman, S.J. Greenberg, B.J. Poiesz, Virology, 186, 619-627 (1992).

23. G.D. Ehrlich, S.J. Greenberg, and M.A. Abbott, PCR Protocols: A Guide to Methods and Applications. A Laboratory Manual, M. Innis, D. Gelfand, J. Sninsky, T. White, eds., Academic Press, Orlando, (1990) pp. 325-336.

24. E.S. Jaffe, W.A. Blattner, D.W. Blayney, P.A. Bunn, J. Cossman, M. Robert-Guroff, and R.C. Gallo, Am. J. Surg. Path., 8, 263-275 (1984).

25. T. Yoshida, M. Kikuchi, K. Ooshima, M. Takeshita, N. Kimura, M. Kozuru, and M. Satoh, Cancer 64, 2515-2524 (1989).

26. M. Shimoyama, Br. J. Haematol. 79, 428-439 (1991).

27. F. Kawano, K. Yamaguchi, H. Nishimura, H. Tsuda, and K. Takatsuki, Cancer 55, 851-856 (1985).

MANAGEMENT OF CYTOMEGALOVIRUS INFECTION IN SOLID-ORGAN TRANSPLANT RECIPIENTS

Brian M. Murray, M.D.
Department of Medicine, State University of New York at Buffalo,
Erie County Medical Center, Nephrology Division
Buffalo, New York 14215

INTRODUCTION

Cytomegalovirus, which gets it's name from the characteristic enlarged cells with intranuclear and cytoplasmic inclusions seen in infected tissue, is a double-stranded DNA virus of the herpes family. After entering a host cell, the viral DNA becomes incorporated into the host genome and the cell is redirected to the production of viral DNA and proteins (1). Most people acquire infection by casual person-to-person contact in childhood and at most experience a mild "flu-like" illness so that about 80% of adults exhibit serological evidence of past infection. As with other viruses of the herpes family the virus can then enter a prolonged period of latency with subsequent reactivation during intercurrent illness or periods of immune suppression. The virus can also be transmitted by blood transfusion, as well as bone marrow or solid organ transplantation. Presumably cells of the transplanted organ may contain CMV DNA in the latent form. The specific cell type responsible is not known with certainty(2).

Cytomegalovirus infection continues to be the most common and serious viral infection complicating transplantation. One of the principal risk factors for posttransplant CMV infection is the serological status of donor and recipient. Infection may represent *primary* infection in a seronegative recipient receiving a graft from a seropositive donor or a *reactivation* of latent infection in a seropositive recipient. As there are many different strains of the CMV virus, seropositive recipients of grafts from another seropositive donor are also at risk of *superinfection* by the donor strain. Indeed such individuals have been shown to excrete both strains simultaneously in the urine(3).

CLINICAL MANIFESTATIONS OF CMV INFECTION

In general, clinical manifestations and the severity of infection is greatest in primary infections though clinical manifestations may range from asymptomatic infection to a severe life-threatening syndrome of multisystem organ failure(4). Patients usually present around 4-6 weeks posttransplant with an unexplained febrile illness characterized by high spiking fevers with generalized arthralgias and myalgias. In many milder cases these clinical symptoms may resolve without therapy but in others can progress to more serious manifestations including a *CMV syndrome* with persistent febrile illness often with *leukopenia* and occasionally *thrombocytopenia*. This in turn may progress to tissue-invasive disease and in many cases the transplanted organ is often disproportionately affected. *CMV-related hepatitis* is characterized by elevated transaminases and is usually most severe in liver transplant recipients (5). Subtle abnormalities of pulmonary dysfunction can be found in virtually all infected individuals (6) but the more severe manifestation of *pneumonia* with a bilateral symmetrical interstitial and alveolar process is fortunately more rare (7). In cardiac transplant recipients, CMV has been linked to both acute rejection and accelerated atherosclerosis (8) and in the kidney a role for CMV has been proposed in causing a distinctive form of *glomerulopathy* though this remains controversial (9,10). Clinical manifestations of CMV can also be seen in the gut with *colitis,* characterized by focal ulceration the most frequent manifestation, though virtually any portion of the gut from mouth-to-anus can be affected (11). Finally, hematogenous spread of the virus may result in ocular involvement with sight-threatening *chorioretinitis* which is usually recognized on funduscopy as a perivascular yellow-white retinal infiltrate (12).

Cytomegalovirus infection also appears to be immunosuppressive in it's own right, and can predispose to the development of other superinfections such as Pneumocystis pneumonia and fungemia. This appears to be due to an effect of the virus itself to impair cellular immunity (13). It has also been speculated that CMV infection, perhaps through up-regulation of cell surface molecules may trigger renal allograft rejection. However, because of the frequency of CMV infection and the difficulty of establishing precisely the onset of viral activation this has been difficult to establish definitively (14). Either way posttransplant CMV infection is associated with significantly decreased allograft and patient survival rates (15,16). The costs of CMV infection are thus considerable both to the patient and to the Health Care system. A recent study from Canada (17) showed that patients who develop CMV disease spend more time in hospital (59 vs. 22 days)and incur considerably increased institutional costs (\$42,611 vs. \$17,309) during the first year posttransplant than their unaffected counterparts. This emphasizes the importance of attempts to prevent, diagnose and effectively treat this important cause of posttransplant morbidity.

PREVENTION OF CMV INFECTION

One possible approach to minimize the incidence of severe CMV disease would be to avoid the use of organs from seropositive donors in seronegative recipients. While Ackerman et al (18) reported a decrease in graft loss and CMV related mortality when they adopted such a policy, Shackleton et al (19) were unable to duplicate these findings. In the absence of a definitive survival advantage this approach is not felt to be justified currently since it would greatly limit the donor pool for such individuals. However it is considered prudent to avoid, whenever possible, the transfusion of CMV-positive blood in such individuals or to use leukocyte-depleted blood products.

Another logical approach would be the pretransplant immunization of seronegative potential recipients. However it must be remembered that in naturally immune individuals, superinfection and serious disease still occur. An attenuated (Towne) strain of the CMV virus has been shown to induce anti-CMV antibody in non-immune dialysis patients awaiting transplant. Trials using the vaccine prior to renal transplantation indicate that although subsequent primary infection is not prevented, the severity of ensuing disease appears to be decreased (20). Thus such an approach has potential utility but would probably not preclude the need for some additional prophylaxis.

PROPHYLAXIS OF CMV INFECTION

Table I lists the agents currently available for use in CMV prophylaxis, the dosages normally employed, and the potential costs. *Passive immunoprophylaxis* has been attempted with both unselected and hyperimmune globulin. CMV hyperimmune globulin is extracted from pooled sera of CMV positive donors and adjusted to a specific titer. Snydman et al.(21) randomized 59 seronegative recipients of a kidney from a seropositive donor to either CMV immune globulin administered intermittently over 16 weeks or no treatment. Although the immune globulin had no effect on subsequent acquisition of infection , it did reduce the severity of infection with significantly less neutropenia, pneumonia and fungal infections in the prophylaxed group. Nevertheless there was still a 21% incidence of CMV disease in the prophylaxed group. CMV immune globulin is FDA approved for the prophylaxis of CMV infection in renal transplantation. Subsequent studies in liver transplant recipients also showed evidence of a decrease in severe CMV-associated disease, though interestingly most of the benefit appeared to accrue to seropositive liver recipients, with no clear-cut benefit in the D+R- group (22). The agent is generally well tolerated with few side-effects and has not been associated with Hepatitis C transmission as occurred recently with some Ig preparations.

Acyclovir is a competitive inhibitor of viral DNA polymerase that can be given orally and is relatively non-toxic but is unfortunately not very active

TABLE I.

CMV PROPHYLACTIC AGENTS AND THEIR COST

AGENT	DOSE	COST
CMV Immune Globulin	150 mg/kg within 72 hr.	$7045
	220 mg/kg weeks 2,4,6 and 8	
	50 mg/kg weeks 12 and 16.	
High-dose Acyclovir	800 mg po qid for 4 months	$1325
Ganciclovir	5 mg/kg iv bid for 30-100d.	$2088(30d)
		$6960(100d)

in vitro against CMV. Balfour et al.(23) reported in a randomized controlled trial a significant reduction in the incidence of posttransplant CMV infection when all patients were prophylaxed with high-dose oral acyclovir for 3 months. The benefit was principally due to a dramatic reduction in CMV disease in the seronegative recipients of a seropositive kidney though the number of patients in this group was small (n=13). All 7 controls but only 1 of 6 acyclovir-treated patients developed CMV disease. Subanalysis of the D+R+ and D-R+ groups showed a trend to less CMV disease that did not reach statistical significance in either, though when all R+ recipients were considered as a single group statistical significance was achieved. Unfortunately the results of other studies have not always reproduced these findings. Bailey et al (24) found that despite high-dose acyclovir, 70% of D+R- recipients developed CMV disease and that this was not helped by addition of (unselected) immune globulin. As a result, high-dose acyclovir is now mostly used as prophylaxis for seropositive renal transplant recipients. These examples illustrate the difficulty in interpreting and comparing many studies on prophylaxis because most are single center studies with considerable heterogeneity of the study populations in terms of serostatus groupings and often insufficient numbers to analyze subgroups, particularly the high-risk D+R- group.

Ganciclovir, another DNA polymerase inhibitor, is much more active against CMV than acyclovir but until recently had to be given intravenously and has potential renal and bone marrow toxicities. Because of the residual incidence of CMV disease in renal transplant patients prophylaxed with acyclovir and CMV immune globulin, and their relative lack of demonstrable efficacy in the more aggressively immunosuppressed patients receiving heart, liver or bone marrow transplants, ganciclovir was the next logical agent to employ in prophylaxis. Initial reports in heart (25) and bone-marrow (26)recipients suggested that ganciclovir begun at the time of transplant and continued for up to 28 (heart) or 100 (bone-marrow) days decreased the incidence of CMV-related illness. Similar results were recently reported for a 100 day course of ganciclovir in liver transplant recipients (27). Finally, Martin et al (28) showed that a similar reduction in CMV disease (9% vs. 28%) could be achieved in liver transplant recipients with just a 14 day course of ganciclovir followed by high-dose oral acyclovir for 3 months. This regimen has the advantage that the ganciclovir can be given during the initial hospitalization and there was not a need for prolonged central vein catheter placement with it's attendant risks. In all 3 studies the major benefit appeared to accrue largely to the seropositive recipients. There have been few studies that have addressed the role of ganciclovir prophylaxis in renal transplant recipients. Dunn et al.(29) tried combining a shorter course of ganciclovir (just 7 days posttransplant) supplemented by 3 doses of CMV immune globulin in a group of solid organ (mostly kidney) transplants, but found no improvement over the results achieved with high-dose acyclovir alone.

Thus there remains considerable variety in the range of prophylactic regimens practiced today. Because of its cost, high-dose acyclovir for 3-4 months posttransplant remains one of the most popular, particularly in seropositive recipients. This should probably be supplemented by intravenous ganciclovir either initially for 7-14 days as utilized by Martin (28) or during antilymphocyte antibody therapy, a particularly attractive and cost-effective form of preemptive therapy or targeted prophylaxis which was recently shown by Hibberd et al.(30) in a randomized controlled trial of seropositive kidney transplant recipients, to be very effective in preventing CMV disease in this cohort. More prolonged initial courses of ganciclovir, with or without immune globulin, appear necessary in the more aggressively immunosuppressed patients such as those receiving intrathoracic or bone-marrow transplants.

The seronegative recipient of a transplant from a seropositive donor remains the greatest challenge. Even though they do appear to reduce the severity of clinical infection, current prophylactic regimens involving either high-dose acyclovir or CMV specific immune globulin or short-course (<30 days) ganciclovir still result in significant residual incidences (20-50%) of symptomatic CMV infection requiring therapy and occasional severe disease. Furthermore the last few years has seen the emergence of new and more potent immunosuppressives such as FK506 and mycophenolate mofetil, with the promise of more to come (31) and it is not clear if these prophylactic regimens will continue to be effective in the new era of immunosuppression that is dawning. There is clearly a need for new ,preferably orally absorbed and in particular more viricidal agents. Recently an oral form of ganciclovir has been developed and although it achieves much lower blood levels of ganciclovir than when the drug is given intravenously it has proved helpful in the management of recurrent CMV retinitis in AIDS (32). Studies are currently underway to evaluate the efficacy of this agent in the prevention of primary and secondary (reactivation) CMV disease posttransplant and it could also have a role to play in the management of recurrent CMV. As outlined earlier CMV infection, when symptomatic, manifests as a febrile syndrome with or without tissue invasive disease. The results of such studies are eagerly awaited though caution is also warranted because of the potential for emergence of ganciclovir-resistant strains. Experience from treatment of HIV infected individuals suggest that this may occur after 3-4 months of continuous ganciclovir treatment(33).

TREATMENT OF SYMPTOMATIC CMV INFECTION

As outlined earlier, CMV infection when symptomatic manifests as a febrile syndrome with or without tissue invasive disease. Ganciclovir is approved by the FDA for treatment of CMV retinitis but felt to be indicated when there is evidence of other tissue invasive disease, eg., pneumonia, enteritis or severe hepatitis (34). No randomized controlled trials have addressed its use for such indications in organ transplants. However, untreated, these conditions carry a very high mortality and several studies, which have reviewed large cohorts of patients treated with ganciclovir for invasive disease, show appreciable reductions in mortality compared to historical controls (34-37), so that randomized trials of ganciclovir therapy for invasive CMV disease at this point would not be indicated, or indeed considered ethical. The efficacy of treatment is traditionally assessed by the clinical response to therapy since conventional or shell vial blood cultures usually turn negative within a few days of initiating ganciclovir, regardless of the clinical course. (38). Van den Berg et al (38), on the other hand, found that the level of CMV antigenemia correlated well with the clinical response. Our experience has been similar. In a cohort of eight patients whom we started on conventional doses of ganciclovir for CMV infection, although five responded promptly with decreased antigenemia, the other three exhibited a continued rise in CMV-Ag with persistent or worsening symptoms but promptly responded when the ganciclovir dose was empirically increased. All three belonged to the group of four patients who had calculated creatinine clearances of < 50 ml/min. This raises the question whether the dosage reductions recommended by the manufacturer for this group of patients resulted in adequate ganciclovir levels. Alternatively the large intervals between dosage adjustments (Creatinine Clearances of 80 , 50 ,25 , 12.5 ml/min) can result in a patient with a clearance of 52 ml/min receiving a dose twice as high as one with a clearance of 48 ml/min . Measurement of ganciclovir levels is not widely available at this time and it is not currently known if the approach of monitoring levels, as is done for aminoglycoside therapy, would increase the clinical response rate and/or reduce the incidence of toxicity. However, serial CMV-Ag counts can be useful in monitoring the effect of ganciclovir therapy and guiding the need for increased dosage.

The role of combining CMV specific immune globulin with ganciclovir in the treatment of invasive CMV disease is controversial (39). Dunn et al (35) reviewing the experience at the University of Minnesota reported excellent results with a 30-day cure rate of almost 90% with ganciclovir alone. On the other hand, George et al (40), in a randomized study suggested that combination therapy of CMV pneumonia may result in fewer ventilator days compared to ganciclovir alone and there was also a trend toward improved long term survival. Combination therapy may be particularly required in the bone marrow transplant recipient where early experience with ganciclovir alone was disappointing (41) but two subsequent studies (42,43) showed improved outcomes with combination therapy though neither included a control group receiving ganciclovir alone. Because of the significant risk of mortality in invasive CMV disease, we favor the combined use of these agents in this setting though a multicenter randomized trial is clearly required to definitively resolve this issue. This is particularly so since the cost of the most commonly recommended regimen (CMV immune globulin, 100mg/kg qod for 7 doses) is considerable.

With the advent of new diagnostic techniques, CMV infection is now often diagnosed at a time when the patient is exhibiting the CMV syndrome (fever and myalgias with or without neutropenia). Although this syndrome can resolve without specific treatment, most practitioners will begin antiviral therapy as soon as the diagnosis is confirmed because of the risk of progression to invasive disease and the evidence that the earlier it is instituted the more successful antiviral therapy proves to be. In this setting most centers will use ganciclovir alone, given the additional cost concerns of immune globulin therapy, which can be reserved for the patients who do not respond to ganciclovir and/or progress to invasive disease. Therapy is generally well-tolerated though occasionally neutropenia can develop or worsen. Our practice is to hold ganciclovir if the absolute neutrophil count (ANC) falls to <1000 and give granulocyte colony stimulating factor, reinnstituting ganciclovir once the ANC rises above 1500.

Because of its tendency to cause nephrotoxicity, there is only anecdotal experience with the use of foscarnet in the CMV infected transplant recipient. In most cases it is reserved for the patient who is progressing despite ganciclovir/immunoglobulin, particularly if there is evidence of persistent antigenemia or the suspicion of ganciclovir resistance which has been found to emerge in patients receiving ganciclovir for prolonged periods.

RELAPSING CMV INFECTION

Although many patients will respond clinically to ganciclovir or combined therapy of CMV infection most reported series report a significant incidence of recurrent CMV infection/disease ranging between 25%-50% (35,44).

This has also been our experience and Figure 1 shows a typical case occurring in a seropositive recipient of a kidney from a seropositive donor. After initially responding to antiviral therapy, symptomatic CMV recurred on two occasions accompanied by a recurrent rise in antigenemia and required two further more prolonged courses of antiviral therapy before ultimate clinical resolution. It appears that recurrent disease may in fact be the result of an inadequate response of the host immune system. Van den Berg et al (44) have shown that patients who recur develop lower levels of lymphocytes carrying the CD8 marker in response to viral infection, while Manez et al have linked the development of recurrent CMV with an atypical type of IgG response to infection (45). The optimal method of managing such patients or of preventing recurrence is unclear. Recurrent disease is much more common in patients receiving a kidney from a seropositive donor and consideration could be given in such patients to following initial IV ganciclovir therapy with a supplemental 6-8 week course of high dose oral acyclovir (800 mg qid PO adjusted for renal function). The development of recurrent symptomatic infection generally requires reinstitution of intravenous ganciclovir. We are currently evaluating the efficacy of oral ganciclovir (32) in this setting.

PREEMPTIVE THERAPY OF CMV INFECTION

A number of new laboratory techniques are available which can even detect evidence of activated CMV virus in the blood before clinical symptoms develop. This has opened the possibility of a third approach to posttransplant CMV infection, which Rubin (45) christened "preemptive therapy" in a 1991 editorial. Preemptive therapy is a form of targeted prophylaxis in which prophylaxis is withheld and antiviral therapy instituted when laboratory evidence of presymptomatic viral infection is found. The Rubin editorial was written to accompany publication of a study (47) in which bone marrow recipients were routinely subjected to bronchoalveolar lavage between 35 and 49 days posttransplant and treated with ganciclovir (5mg/kg bid for 2 weeks, then 5 days/week until day 120), if evidence of CMV was found on cytologic studies or shell-vial culture of aspirate. Control patients were observed sequentially and treated only if CMV pneumonia developed. Prophylactic ganciclovir reduced the incidence of CMV pneumonia from 70 percent in the control group to 25%.

There are now several other candidate tests that could be used for guiding institution of such therapy - the shell vial blood culture, and/or the detection of CMV antigenemia by monoclonal antibodies or viral DNA by PCR in white cells isolated from peripheral blood (48). What is needed is a test that not only detects evidence of virus sufficiently in advance of symptoms but also is predictive of subsequent development of CMV infection and the need for antiviral therapy. Singh et al. (49) compared in a randomized fashion a

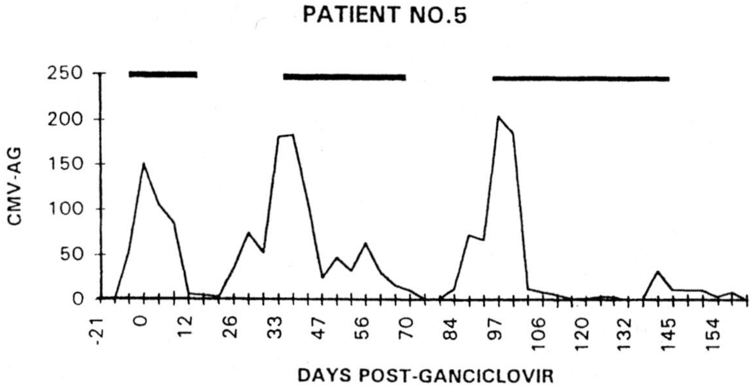

FIGURE I

Renal transplant patient with relapsing CMV infection.

preemptive short (7 day) course of intravenous ganciclovir against conventional high-dose acyclovir prophylaxis in liver transplant recipients. Patients were monitored weekly by shell-vial blood culture and preemptive therapy given as soon as virus was detected. Preemptive ganciclovir was well tolerated and significantly reduced subsequent viral shedding and CMV disease. However, most (>85%) of patients fell into the seropositive category and only four were D+R-, an insufficient number to conclude whether this particularly high risk group would benefit.

In a prospective study of 36 renal transplant recipients (50), we compared the ability of the Shell vial and CMV antigen tests to predict subsequent need for ganciclovir therapy. Firstly, we evaluated the utility of several degrees of antigenemia in identifying patients for preemptive therapy. A **CMV-AG count of >50/50,000 PBLs** was used as the criterion for institution of preemptive therapy during the study because of previous evidence from our group that this predicted a high likelihood of subsequent CMV disease (51) This led to 8 of the 10 patients with symptomatic CMV receiving antiviral therapy. However treatment was not truly preemptive in that most already were symptomatic, albeit mildly. If we had used the criterion of a positive CMV-AG test to institute therapy this would have resulted in treatment of an additional 15 patients most of whom were never symptomatic. If the object is for treatment to be truly preemptive and to include as many subsequently symptomatic patients as possible then our study would suggest that a suitable threshold might be a **CMV-AG count >10**. Use of this threshold would have resulted in earlier institution of therapy by about 4 days in the 8 patients treated and the two additional patients would have qualified for therapy. Both subsequently developed mild clinical symptoms and their CMV-AG peaked at 34 and 48 respectively but then fell spontaneously to low levels as symptoms resolved. Use of a positive **blood culture (Shell Vial)** as the criterion would have delayed therapy by 4 days in our 8 patients and would have resulted in treatment of an additional 8 patients, 6 of whom were never symptomatic. We found that if a threshold value of CMV-Ag of 10/50,000 PBL's was used, this had a positive predictive value of 100% for subsequent symptomatic CMV infection, whereas use of a positive Shell vial would result in "unnecessary" therapy of as many as 25% of at risk patients.

We recently initiated a randomized trial of preemptive therapy in at risk renal transplant recipients (seropositive recipients and seronegative recipients of a kidney from a seropositive individual). The PREEMPTIVE group were randomized to receive serial weekly monitoring with the CMV-AG test combined with institution of combined antiviral therapy with ganciclovir (21 days) and seven doses of CMV specific immune globulin every other day. The CONTROL group received prophylaxis with either acyclovir for 3 months (seropositive recipients) or CMV specific immune globulin (seronegative recipients of a seropositive kidney) and antiviral therapy was deferred until the development of symptomatic CMV infection with laboratory confirmation of infection. Preliminary analysis suggests that the preemptive approach significantly decreases the subsequent incidence of symptomatic CMV infection and there was also a trend to less tissue invasive disease. Similar numbers (≈ 50%) in each group required at least one additional course of antiviral therapy for recurrent antigenemia or symptomatic CMV. Most of the problems with recurrent CMV occurred in the D+R- group (of whom there were 6 patients in each group), who exhibited a 40% requirement for repeat antiviral therapy. It may be that early and aggressive prophylaxis (e.g. initial intravenous followed by prolonged oral ganciclovir) either at the time of transplant or on detection of antigenemia, may be necessary to more effectively prevent relapsing disease in this group. This could then be supplemented with monitoring for CMV antigenemia or DNAemia by PCR with reinstitution of intravenous drug if a rising titer of antigenemia is found.

Thus, while significant advances in the prevention and management of posttransplant CMV have been made, there remains much work to be done. In particular better and more viricidal agents, which can be given orally are needed and the questions of universal prophylaxis versus targeted preemptive therapy and single versus combination therapy of established disease will need to be resolved. Larger multicenter trials will be needed, particularly in the very high-risk D+R- patients, preferably stratified by the type of transplant since it is to be expected that more extensive prophylaxis/treatment may be needed in the more heavily immunosuppressed individuals receiving bone marrow or heart-lung than for example, those receiving kidney transplants. Finally such trials will also have to address in detail the question of cost-effectiveness, since in these days of managed care, sometimes the price of success can be too high.

REFERENCES

1. P.D. Griffiths and J.E. Grundy. Biochem J, 241, 313-324 (1987).

2. J.W. Gnann, J. Ahlmen, C. Svalander, L. Olding, M.B.A. Oldstone and J.A. Nelson. Am J Pathol, 132, 239-248 (1988).

3. S. Chou. N Engl J Med, 314, 1418-1423 (1986).

4. R.L. Simmons, A.J. Matas and L.C. Rattazzi,et al. Surgery, 82, 537-546 (1977).

5. C.V. Paya, P.E. Herman and R.H. Wiesner, et al. J Infect Dis, 160, 752-758 (1989).

6. W.J. Van Son, W. van der Bij and T.H. The, et al. Transplant Proc, 21, 2065-2068 (1989).

7. J.E. Craighead JE. Pathbiol Annu, 102, 691-703 (1970).

8. M.T. Grattan, C.E. Moreno-Cabral and V.A. Starnes,et al. JAMA, 261, 3561-3566 (1989).

9. W.P. Richardson, R.B. Colvin and S.H. Cheeseman, et al. N Engl J Med, 305, 57-63 (1981).

10. G.A. Herrera, R.W. Alexander and C.F. Cooley, et al. Kidney Int, 29, 725-733 (1986).

11. R.W. Goodgame. Ann Intern Med 119, 924-935 (1993).

12. J.N. Bloom and A G Palestine. Ann Intern Med 109, 963-969 (1988).

13. W.P. Carney, R.H. Rubin, R.A. Hoffman, W.P. Hansen, K. Healey and M.S. Hirsch. J Immunol 126, 2114-2116 (1981).

14. S.L. Simmons, R. Weil and M.B. Tallent, et al. Transplant Proc 2, 419-423 (1970).

15. S.C. Marker, R. Howard, R. Simmons, J.M. Kalis, D.P. Connelly, J.S. Najarian and H.H.Balfour, Jr. Surgery, 89, 660-671 (1980).

16. R.M. Lewis, P.C. Johnson, D. Golden, C.T. Van Buren, R.H. Kerman and B.D. Kahan. Transplantation, 45, 353-359 (1988).

17. J.M. Mc Carthy, M.A. Karim, H. Krueger and P.A. Keown. Transplantation, 555, 1277-1282 (1993).

18. J.R. Ackerman, W.M. LeFor and S. Weinstein, et al. Transplant Proc, 20, (1,Suppl.1) 469-471 (1988).

19. C.R. Shackleton, P.A. Keown, Landsberg DN, et al. Transplant Proc,.23, (1, pt 2) 1350-1351 (1991).

20. S.A. Plotkin, R. Higgins, JB Kurtz, et al. Transplantation, 58: 1176-1178 (1994).

21. D.R. Snydman, B.G. Werner, B. Heinze-Lacey B, et al. N Engl J Med, 317,1049-1054 (1987).

22. D.R. Snydman, B.G. Werner, N.N. Dougherty, et al. Ann Intern Med 119, 984-991 (1993).

23. H.H. Balfour, B.A. Chace, J.T. Stapleton, R.L. Simmons and D.S. Fryd. N Engl J Med, 320, 1381-1387 (1989).

24. T.C. Bailey, N.A. Ettinger, G.A. Storch, et al. Am J Med, 95, 273-278 (1993)

25. T.C. Merigan, D.G. Renlund, S. Keay, et al. N Engl J Med, 326, 1182-1186 (1992).

26. J.M. Goodrich , M. Mori, C.A. Gleaves, et al. N Engl J Med, 325, 1601-1607 (1992).

27. D.J. Winston, D. Wirin, A. Shaked and R.W. Busuttil. Lancet, 346, 69-74 (1995).

28. M. Martin, R. Manez, P. Linden, et al. Transplantation, 58, 779-785 (1994).

29. D.L. Dunn, K.J. Gillingham, M.A. Kramer, et al. Transplantation, 57, 876-884 (1994).

30. P.L. Hibberd, N.E. Tolkoff-Rubin, D. Conti D, et al. Ann Intern Med, 123, 18-26 (1995).

31. R. Calne. Transplant Proc, 24, 1260-1262 (1992).

32. W.L. Drew, D. Ives, J.P. Lalezari, et al. N Engl J Med, 333, 615-620 (1995).

33. D.G. Wolf, I.L. Smith, D.J. Lee, W.R. Freeman, M. Flores-Aguilar and S.A. Spector. J Clin Invest, 95, 257-263 (1995).

34. D.R. Snydman. Rev Infect Dis, 10, (Suppl. 3) s554-s562 (1988).

35. D.L. Dunn, J.L. Matoral, K.J. Gillingham, et al. Transplantation, 51, 98-106 (1991).

36. C.V. Paya, P.E. Hermans, T.F. Smith, et al. Transplantation, 46, 229-234 (1988).

37. R. Hrebinko, M.L. Jordan, J.S. Drummer, et al. Transplant Proc, 23, 1346-1347 (1991).

38. A. P. Van den Berg, A.M. Tegzess, A. Scholten-Sampson, et al. Transplant Int, 5, 101-106 (1992).

39. R.H. Rubin RH. Kidney Int, 44, 221-236 (1993).

40. M.J. George, D.R. Snydman, B.G. Werner, et al. Transplant Proc, 25, (Suppl4) 22-24 (1993).

41. I. Frank and H.M. Friedman. Ann Intern Med 109, 769-771 (1988).

42. E.C. Reed, R.A. Bowden, P.S. Dandliker, K.B. Lilleby KB and J.D. Meyers Ann Intern Med 109, 783-788 (1988).

43. D. Emanuel, I. Cunningham, K. Jules-Elysee, et al. Ann Intern Med, 109, 777-782 (1988).

44. A.P. Van den Berg, W.J. Van Son, E.B. Haagsma, et al. Transplantation, 55, 847-851 (1993).

45. R. Manez, M.K. Breinig, S. Kusne, et al. Transplantation, 59, 1220-1223 (1995).

46. R.H. Rubin. N Engl J Med, 324, 1057-1059 (1991).

47. G.M. Schmidt, D.A. Horak, J.C. Niland, et al. Engl J Med 324, 1005-1011 (1991).

48. T.H. The. Transplantation, 54, 193-198 (1992).

49. N. Singh, V.L. Yu, L. Mieles, M.M. Wagener, R.C. Miner, T. Gayowski. Ann Intern Med 120, 375-381 (1994).

50. B.M. Murray, D. Amsterdam, V. Gray, J. Myers J and R.C. Venuto. J Am Soc Nephrol (Submitted for Publication).

51. B.M. Murray, J. Brentjens, D. Amsterdam, J. Myers, V. Gray, J.P. Singh and R.C. Venuto. J Am Soc Nephrol 4, 1615-1622 (1994).

PART IV

SYNDROMES OF VARIOUS INFECTIOUS ORIGINS

IMMUNODIAGNOSIS OF PRION DISEASE

R.J. Kascsak, R. Fersko, D. Pulgiano, R. Rubenstein and R.I. Carp
NYS Institute for Basic Research in Developmental Disabilities
Staten Island, New York 10314

ABSTRACT

The immunodiagnosis of prion diseases is of critical importance due to the transmissibility of these conditions and their fatal prognosis. A panel of monoclonal and polyclonal antibodies have been generated for use in the study and diagnosis of these diseases. This manuscript describes the generation and characterization of these antibodies as well as their diagnostic application.

INTRODUCTION

Prion diseases are transmissible disorders of the central nervous system whose parameters are controlled by strong genetic influences (1). These neurological conditions occur in a wide range of mammalian genera including humans (2). Scrapie in sheep and goats is the prototype of prion diseases and this form of disease has been known for over two hundred years. The human prion conditions include Creutzfeld-Jakob disease, Gerstmann-Sträusler syndrome, kuru and fatal familial insomnia. Recently this group of disorders has spread to new hosts including cattle and domestic cats. While these conditions occur at a low frequency, diagnosis is important because of their transmissibility and fatal prognosis. The hallmark of these disorders is the conversion of a host membrane glycoprotein termed PrP^c {or PrP^{sen}} into a protease resistant highly aggregated form termed PrP^{Sc} { or PrP^{res}} (3). Mutation in the gene which codes for this protein in the form of point mutations, insertions and deletions are associated with genetically inherited forms of these diseases (4). The majority of human prion diseases (greater than 95% of cases) occur sporadically with no known source of infection. The only known contributing factor is a higher incidence of disease associated with homozygosity for a polymorphism at codon 129 of the PrP gene (5).

The rapid and accurate diagnosis of prion diseases is of critical importance to the medical communities. The immunological detection of PrP using either monoclonal or polyclonal antibodies to PrP affords the sensitivity and specificity required for such an assay.

The basis of the humoral immune response to PrP will be discussed as well as the utilization of antibody based assays for the diagnosis of prion disease in a wide range of animal species.

MATERIALS AND METHODS

PREPARATION OF PRP ANTIGEN: PrP^{Sc} was isolated from the brains of mice or hamsters infected with scrapie agent (6). The preparation of the inoculum, injection, scoring and sacrifice of the animals were performed as previously described (6). PrP^{Sc} was purified from brains of infected animals at the clinical stage of disease by a modification of the procedure of Hilmert and Diringer (7) as previously described (6). The procedure involved the use of detergent extraction, differential centrifugation, and treatment with proteolytic enzymes. The final pellet obtained from 12 hamster brains or 24 mouse brains contained 100 to 500 μg of PrP^{Sc} and 1 to 10% of the infectivity found in the original homogenate (6).

IMMUNIZATION: Animals (Balb/CJ mice, Syrian LVG/LAK hamsters, New Zealand white rabbits or PrP knockout mice (8)) were immunized with non-infectious formic acid extracted PrP as previously described (9). Monoclonal antibodies were produced from Balb/CJ or RBF/Dn mice immunized with hamster PrP. Clones produced from such fusions yielded monoclonal antibodies 3F4 and 7G5(6). Animals were immunized with 20-40 μg of purified PrP^{Sc} which had been solubilized and rendered non-infectious using formic acid (9). Antigen was emulsified in Hunter's titer max (Vaxcel,Norcross,GA) and each animal receiving 3-4 immunizations. Following such a regimen, rabbits were used to produce polyclonal antibody and mice used to generate monoclonal antibody.

IMMUNOASSAYS: ELISA, Western blot immunofluorescence and immunohistochemical assays were carried out as previously described (9) and used to assay immune response to PrP. In addition, the immunodiagnosis of prion disease using Western blot and immunohistochemistry was performed using these reagents.

PEPSCAN ANALYSIS: Synthesis of a complete set of overlapping solid-phase dodecapeptides (12 peptides) on polyethylene rods and the subsequent detection by ELISA assay were carried out according to established PEPSCAN procedures (10). For the synthesis of peptides, the sequence data of mouse PrP determined by Locht et al. (11) were used. Peptides were considered to represent antigenic sites when peaks occurred in a set and at least one of the peaks in such a set reproducibly attained a value more than twice the background.

RESULTS

The immunodiagnosis of prion diseases is dependent upon the availability of highly specific antibodies to PrP which react with this protein originating from a wide range of mammalian sources. The goal of our laboratory has been to generate a variety of polyclonal and monoclonal antibodies to PrP and to apply such reagents for research and immunodiagnosis. Antibodies to PrP have been produced in three different animal species (rabbits, mice and hamsters) (Table 1). The ability to generate such a response appears to be based upon self-nonself recognition i.e. mice are tolerant to mouse PrP while immunoreactive to hamster PrP. In addition PrP knockout mice, which are devoid of indigenous PrP, produce an avid humoral response to foreign mouse PrP. Rabbits responded to four major epitopes on mouse PrP ; such epitopes are absent from the animal which generated the immune response. Interestingly inter-species relationship among PrP from various sources become evident using this approach. Antibodies generated in rabbits to mouse PrP react with all mammalian species except ferrets, mink and rabbits. The reactive epitopes are mainly located in the C-terminal region of the protein, amino acids 180-225. The response in mice and hamsters , as seen in other species, is self-nonself dependent. Animals will not generate an immune response to homologous PrP. The region between amino acids 109-114 appears to be an important antigenic site. The antibody response to mouse PrP generated in hamsters is the only antibody known to immunoreact with rabbit and mink PrP. Monoclonal antibody 3F4 recognizes an epitope present on hamster and human PrP between amino acids 109-113 which is dependent upon the presence of methionines at positions 109 and 112. An additional monoclonal antibody (7G5) has been produced in response to hamster PrP and this antibody reacts exclusively with hamster PrP(6). The immune response to mouse PrP in the single PrP knockout mouse examined thus far is directed toward the same epitope as that seen in rabbits to the same antigen. Another interesting inter-species relationship is seen in the response in mice to PrP peptide 176-184. The antibodies generated reacted with all PrP species examined except ferret and rabbit, both of which differ from the other species at either position 177 (leucine to phenylalanine in ferrets) or position 180 (leucine to aspartic acid in human).

The species specific recognition of PrP by these different antibodies is shown in Table 2. The immunoreactivity of the different PrPs is based on shared epitopes and reflect the genealogy of PrP. Avian PrP has been sequenced and retains approximately a 50% homology with mammalian PrP (12). Despite this homology there appears to be no shared epitopes between the antibodies generated to these different PrPs. Antibody generated to avian PrP does not crossreact with mammalian PrP and likewise antibodies to mammalian PrPs do not crossreact with avian PrP. Arachnids (hay mites) were also investigated for the presence of PrP immunoreactive protein using this panel of antibodies. Several antibodies to PrP (see Table 2) crossreact with protein(s) present in

TABLE 1

PEPSCAN* ANALYSIS OF ANTIBODY TO PrP

Antibody	Antigen	Position in PrP	Amino Acid Sequence	Reactivity
Rabbit 78295	Mouse PrP	99-105 181-189 201-207 220-224	NKPSKPK TIKQHTVT DVKMMER SQAYY	All except Ft, Mk, Rb
Hamster C3	Mouse PrP	110-114	KHVAG	Bv, Fe, Ft, Mo, Mk, Rb, Rt, Sh
Mab 3F4	Hamster PrP	109-113	MKHM	Fe, Ha, Hu
Mab 7G5	Hamster PrP	163-171	YRPVDQYN	Ha
Knockout 94-5	Mouse PrP	201-207	DVKMMER	similar to 78295
Mouse	FF1 Peptide	176-184	NFVHDCVN	all except Ft, Rb

*Analysis performed by Dr. J. Langeveld, Institute for Animal Science and Health, Netherlands.
Bv, bovine; Fe, feline; Ft, ferret; Ha, hamster; Hu, human; Mk, mink; Mo, mouse; Rb, rabbit, Rt, rat; sh, sheep.

TABLE 2

ANTIBODY REACTIVITY WITH PrP^c FROM VARIOUS SPECIES

	Bov	Fel	AHa	CHa	SHa	Hu	Mk	Mo	Rb	Rt	Sh	Chk	ARA
Mab3F4	-	+	+	-	+	+	-	-	-	-	-	-	-
Mab7G5	-	-	+	+	+	-	-	-	-	-	-	-	-
rabbit anti-MoPrP	+	+	+	+	+	+	-	+	-	+	+	-	+
rabbit anti-SHaPrp	+	+	+	+	+	+	-	+	-	+	+	-	+
rabbit anti-ChkPrP	-	-	-	-	-	-	-	-	-	-	-	+	+
hamster anti-MoPrP	+	+	-	+	-	-	+	+	+	+	+	-	+
PrP Knockout mouse	+	+	+	+	+	+	nd	+	-	+	+	-	-

ARA, arachnid; Bov, bovine; Chk, chicken; Fel, feline; AHa, armenian hamster; CHa, chinese hamster, SHa,syrian hamster; Hu, human; Mk, mink; Mo, mouse; Rb, rabbit; Rt, rat; Sh, sheep.
nd-not done.

mite preparations. The sequence homology of this immunoreactive protein to mammalian or avian PrP is being investigated.

Antibodies generated to PrP have been used to examine the cellular distribution of this protein and as diagnostic aids to study prion disease. Immunofluorescence studies have revealed the topographical distribution for PrP^c (Figure 1). This protein is linked by a glycosylphosphatidylinositol (GPI) anchor to the surface of all mammalian cells. The concentration of this protein appears to be highest on cells of the central nervous system especially neurons and astrocytes. The distribution of this protein on the surface of a reactive astrocyte is shown in Figure 1. The protein appears to cover the entire surface of such cells. During prion infection this protein is abnormally processed resulting in the acquisition of protease resistance and in its accumulation within lysosomal vesicles. Over time this leads to the extra-cellular accumulation in the form of PrP^{Sc} aggregates and amyloid.

The diagnosis of prion disease is most often accomplished using Western blot analysis or immunohistochemistry for the detection of PrP^{Sc}. Western blot testing is most often performed because diagnosis is based on several parameters of PrP^{Sc} and the results can easily be interpreted based upon three characteristics of PrP^{Sc}. The assay relies on these important criteria; the insolubility of PrP^{Sc} in detergent, conversion to specific molecular weight forms following protease treatment and immunoreactivity with highly specific antibodies to PrP. Several monoclonal and polyclonal antibodies are available for such diagnostic studies. The identification of PrP^{Sc} derived from infected hamster, mouse and human brain tissue is shown in Figure 2. PrP proteins were isolated by centrifugation and electrophoresed with or without prior protease treatment. Proteins were transferred to nitrocellulose and immmunoblotted using highly specific rabbit polyclonal antibody to PrP. The characteristic shift in molecular weight following protease treatment of the immunoreactive proteins assures the correct identification of PrP^{Sc} and the proper diagnosis.

The critical importance for rapid and accurate diagnosis of human prion disease was recently emphasized by an outbreak of CJD in young adults (Table 3). These cases in Great Britain(13,14) and France (15) occurred in patients 16-39 years of age, an age significantly younger than normally observed for such patients. Sporadic CJD normally occurs in patients 50-70 years of age and is not known to occur naturally within this youngerr age group. However, patients of this age have been known to become infected iatrogenically i.e. through the use of contaminated products or instruments. An example

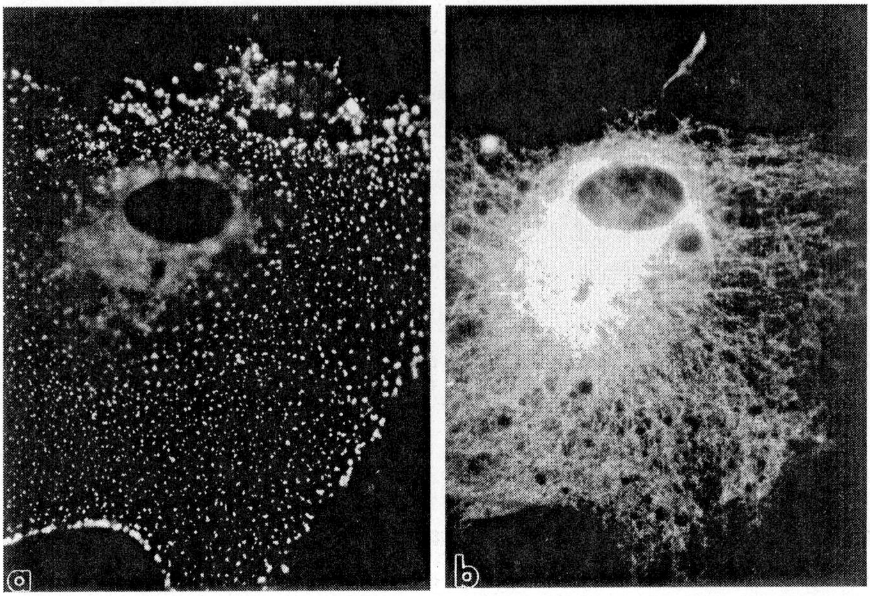

Figure 1
PrP Distribution in Astrocyte Cultures. Astrocytes were cultured from fetal hamster brain and immunostained with antibodies to PrP (A) or glial fibrillary acidic protein (B).

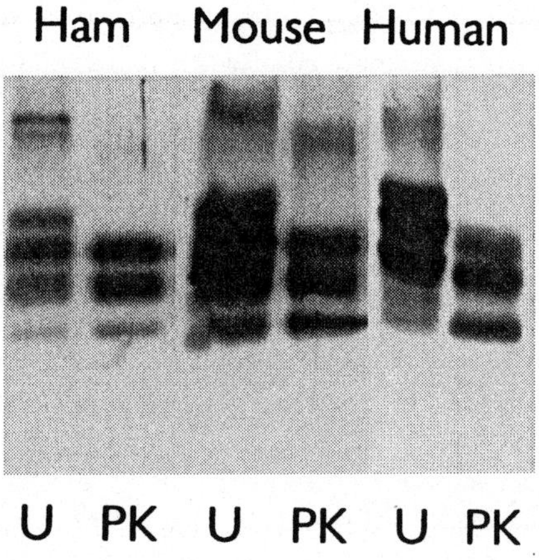

Figure 2
Western Blot Analysis of PrP^{Sc} in Three Different Species. PrP^{Sc} was isolated by differential centrifugation from the brains of prion infected hamsters (Ham), mice or humans. PrP^{Sc} was analyzed by PAGE either untreated (U) or following treatment with proteinase K (PK). Samples were immuno-blotted as previously described (9).

of such contamination was the use of pituitary dergrowth hormone obtained from human tissue (Table 3) (16). For the cases associated with the BSE epidemic there is at present no direct evidence, despite extensive investigation, for transmission between cattle and humans. Nevertheless, even without such direct evidence, the transmission of agent between BSE cattle and humans appears to be the most likely explanation for these unusual cases (13).

DISCUSSION

It has been the goal of this laboratory to generate a wide range of monoclonal and polyclonal antibodies to PrP. Such reagents are important tools in the study and diagnosis of human and animal prion disease. These antibodies provide the sensitivity and specificity only available through the use of such reagents. Such antibodies serve as the resource for the rapid diagnosis of prion disease such as the outbreak in cattle (17) and recently in young adults (13). These reagents have also provided some unique insight into the inter-species relationship among PrP proteins and prion disease. The immunoreactivity of monoclonal antibody 3F4 and polyclonal antibody C3 are mutually exclusive in that they recognize analogous epitopes on different species of PrP (see Table 1). However feline PrP is unique in that it is immunoreactive with both antibodies suggesting an unusual sequence and genealogy for this protein. Recently polyclonal rabbit antibody to PrP has been shown to react with a protein present in arachnid species (18) further supporting the high degree of conservation for PrP protein (19). The association of abnormal forms of this protein with CNS disease in mammalian species is well established. The potential association with similar diseases in avian and lower invertebrate species is certainly possible and intriguing.

The unique nature of prions creates the risk for potential outbreaks in animal and human populations. These risks are further complicated by many unresolved questions surrounding these agents. One of the major issues involves their relationship to other forms of infectious agents. The prion concept suggests that these agents are composed exclusively or almost exclusively of PrP protein (1). The wide diversity of prion properties within a single host containing a single PrP coding sequence has suggested that other factors may play a role. The diversity and complexity of prion would appear to extend beyond the informational content of the primary sequence of PrP protein alone. The role of a small non-coding nucleic acid associated with the prion agent has been suggested as exemplified in the virino theory (1). It has also been suggested that such properties would not have to involve a classic nucleic acid coding sequence but perhaps only an additional protein moiety associated with

TABLE 3

PRION DISEASE IN YOUNG ADULTS

Diagnosis	Number of Cases	Location	Patient Age	Pathology	Reference
Sporadic CJD	10	Great Britain	16-39 mean 27	spongiosis cortical and cerebellar plaques	Will et al. Lancet 347: 921, 1996
Sporadic CJD	1	Great Britain	28	spongiosis cortical plaques	Tabrizi et al. Lancet 347, 941, 1996
Sporadic CJD	1	France	26	spongiosis cortical plaque	Chazot et al. Lancet 347: 1181, 1996
Growth Hormone related CJD	45	US, France Great Britain	12-39	spongiosis cortical plaques	Brown et al. Lancet 340: 24, 1992

PrP which one proposal has been designated protein X (20). Regardless of the nature of the infectious agent, it had been believed that PrP^{Sc} was generated as part of the agent replication process. Recent studies have however demonstrated the in vitro cell free conversion of PrP^{C} into protease resistant PrP^{Sc} (21). Such conversions have only been shown to take place in the presence of pre-existing PrP^{Sc}. It is not known if de novo generation is possible. It is also not known if the newly generated PrP^{Sc} is associated with infectivity.

The true nature of prions remains controversial and the basis of much speculation. Prions may indeed arise from a rare de novo event leading to a protease resistance transmissible form of PrP. This rare event would reflect the incidence of sporadic CJD, two per million in the general population. Others have suggested that prion disease arises in another perhaps equally unique manner. In this scenario, the agent would be ubiquitous but gives rise to disease in only genetically susceptible individuals. Susceptibility might be controlled through the PrP gene itself. Once disease occurs, the disease could be transmitted by direct CNS to CNS contact. Considerable information has come to light in the past few years concerning these agents but many issues remain unresolved. This group of diseases which once mystified the primitive Fore tribe of New Guinea continues to surprise and perplex modern scientists.

ACKNOWLEDGEMENTS

We thank R. Weed and M. Nascimento for photographic assistance, and Pat Calimano for typing the manuscript. We also thank Dr. J. Langeveld, the Netherlands for performing pepscan analysis. This work was supported in part by the New York State Office of Mental Retardation and Developmental Disabilities and NIH grant 1203 AG14162.

REFERENCES

1. Carp, R.I., Kascsak, R.J., Wisniewski, H.M., Merz, P., Rubenstein, R., Bendheim, P. and Bolton, D. In: Alzheimer's Disease and Associated Disorders 3:79-99,1989.

2. Prusiner, S.B. and Hsiao K. Ann. Neurol. 35: 385-395,1994.

3. Meyer, R.K., McKinley, M.P., Bowman, K.A., Brawnfeld, M.B., Barry, R.A. and Prusiner.S.B. Proc. Natl. Acad. Sci. U.S.A. 83:2310-2314,1986.

4. Goldfarb, L. and Brown P. Ann. Rev. Med. 46:57-65,1995.

5. Owen, F., Poutter, M., Collinge, J. and Crow T.J. Am. J. Hum. Genet. 46:1215-1216, 1990.

6. Kascsak, R.J., Tonna-DeMasi, M., Fersko, R., Rubenstein, R., Carp R.I.and Power J.M. In: Brown F. ed. Transmissible Spongiform Encephalopathies-Impact on Animal and Human Health. Dev.Biol.Stand. Basel, Karger. 80: 141-151.1993.

7. Diringer, H., Gelderblom, H., Hilmert, H., Ozel, M., Edelbluth. C.and Kimberlin, R.H. Nature 306:476-478 ,1983.

8. Buder, H., Aguzzi, A., Saller, A., Grelner, R., Autenued, P., Aguet M. and Weissman, C. Cell 73:1339-1347, 1993.

9. Kascsak, R.J., Rubenstein, R., Merz, P., Tonna-DeMasi, M., Fersko, R., Carp, R.I. Wisniewski, H.M.and Diringer, H. J. Virol. 61:3688-3693, 1987.

10. Geysen, H.M., Meloen, R.H. and Barteling, S.J. Proc. Natl. Acad. Sci. U.S.A. 81:3998-4002, 1984.

11. Locht, C., Chesebro, B., Race, R. and Keith, J.M. Proc. Natl. Acad. Sci. U.S.A. 83:6372-6376, 1986.

12. Harris, D., Falls, D., Johnson F. and Fischback, G. Proc. Natl. Acad. Sci. U.S.A. 88:7614-7668, 1991.

13. Will, R.G., Ironside, J.W., Zeidier, M., Cousens, S.N., Estibeiro, K., Alperovitch, A. Pose, S., Pocchiari, M., Hofman, A., and Smith, P.G. Lancet 347-921-925, 1996.

14. Tabrizi, S.J., Scaravilli, F., Howard, R.S., and Collinge J. Lancet 347:945-948, 1996.

15. Chazot, G., Broussolle E., Lapras, C.I., Blattler, T., Aguzzi, and Kopp N. Lancet 347:1181, 1996.

16. Brown, P., Preece, M., and Will R. Lancet 340:24-27, 1992.

17. Scotti, A.C., Wells, G.A., Stack, J.J., White, H., and Dawson, M. Vet. Microbiol. 295-304, 1990.

18. Wisniewski, H.M., Siguardarson S., Rubenstein, R., Kascsak, R.J. and Carp, R.I. Lancet 347:1114, 1996.

19. Westaway, D.and Prusiner, S.B. Nucleic Acid Res. 13:2035-2044, 1986.

20. Telling, G.C., Scott M., Mastrianni, J., Gabizon, R., Torchia, M., Cohen, F., DeArmond, S., and Prusiner, S.B. Cell 83:79-90, 1995.

21. Kocisko, D., Priola, S., Raymond G, Chesebro, B., Lansbury, P. and Caughy B. Proc. Natl. Acad. Sci. U.S.A. 92:3923-3927, 1995.

CHRONIC FATIGUE SYNDROME

A. C. Mawle
Division of Viral and Rickettsial Diseases
Centers for Disease Control and Prevention
Atlanta, Georgia 30333

ABSTRACT

Chronic fatigue syndrome (CFS) has emerged as a public health concern over the past decade. A working case definition was created in 1988 and revised in 1994, and this has been used to establish prevalence estimates using physician-based surveillance and an a random digit dial telephone survey. Although CFS has some characteristics of an infectious disease, so far no infectious agent has been associated with the illness. Studies of immune function in CFS patients failed to detect differences between cases and healthy controls. However, when cases were subgrouped according to whether they had a sudden or gradual onset, differences in immunologic markers were detected between cases and their matched controls.

INTRODUCTION

Chronic fatigue syndrome (CFS) is an illness of unknown etiology which has emerged as an important public health issue over the last decade. It is characterized by debilitating fatigue, accompanied by a variety of other symptoms, including problems with short-term memory, difficulty concentrating, muscle and joint pain, headaches, sore throat, tender lymph nodes, unrefreshing sleep and post-exertional malaise. In 1988, a group of experts convened by the Centers for Disease Control and Prevention (CDC) devised a case definition for CFS in attempt to standardize the patient groups used in CFS research (1). This definition was revised in 1994 by an international working group, including physicians from the United states, Australia, Great Britain, Italy, Sweden, and the Netherlands, to reflect the broader experience with the illness (2).

PREVALENCE OF CFS

In 1989, CDC set up a surveillance system for CFS in four cities in the US in order to establish the prevalence of the illness. The cities were Wichita, Kansas, Grand Rapids, Michigan,

Reno, Nevada and Atlanta, Georgia. All physicians in specialties likely to see CFS were asked to participate by referring patients with unexplained fatigue or chronic unwellness to the surveillance system. If the patient agreed to participate, they were then thoroughly evaluated for CFS (3), using the 1988 case definition. After 4 years, the surveillance was terminated, and the prevalence rates for each of the 4 cities determined. The crude prevalence rates ranged from 3.8 per 100000 in Atlanta to 9.6 per 100000 in Wichita, with an overall rate of 5.2 per 100000 (3).

Since this system relied upon physician referral, the rates generated are necessarily a minimum estimate, so in 1992 we took an alternative approach to estimating the prevalence of CFS. We conducted a random digit dial telephone survey in San Francisco, California and asked if any member of the household had been fatigued for 1 month or longer. Anyone answering in the affirmative was asked to participate in the surveillance survey and answer a detailed questionnaire about their illness. This information was used to characterize respondents as "CFS-like", using the 1994 case definition. These people were not interviewed in person, nor were they seen by a physician to eliminate other potential causes of fatigue, and thus this prevalence estimate will be higher than the true prevalence. We estimated that the prevalence rate of CFS-like illness was 200 per 100000.

IS CFS AN INFECTIOUS DISEASE?

Evidence that CFS might be caused by an infectious agent has come from both epidemiologic and immunologic evidence. A series of outbreaks of fatiguing illnesses have been reported earlier in this century, in particular in Los Angeles, USA (1934), Akureyri Iceland (1948) and the Royal Free Hospital, UK (1955). Many of these outbreaks were thought to be atypical polio, though no causative agent was ever found. This historical link has fueled the search for evidence of enteroviral infections in CFS. In addition, it has been known for some time that chronic infection with a variety of agents, such as the herpesviruses Epstein-Barr virus (EBV) and cytomegalovirus (CMV), enteroviruses, retroviruses, and *Borrelia burgdorferi* can lead to a fatiguing illness. In 1985 an outbreak of a chronically fatiguing illness was reported to be associated with elevated levels of antibody to the early antigen (EA) of EBV. However, this association was not confirmed and subsequent studies by other investigators showed that though some patients with fatiguing illness do have elevated EA titers, it is not useful in diagnosing CFS (4).

Perturbations in the immune system have also been described in CFS. These include increased activation markers (CD38, DR) on CD8 T cells (5), a decrease in the percentage of circulating T suppressor cell population (CD8 CD11b) (5), and a decrease in circulating naive helper T cells (CD4 CD45RA) (6). Low numbers of circulating natural killer cells and decreased NK function have been reported (6-12), as have decreased lymphocyte proliferative responses (13) and low or absent delayed-type hypersensitivity responses as measured by skin testing (14). However, none of these differences are found consistently in all patient groups. These changes, although inconsistently found, are suggestive of the kinds of changes seen in an acute infection, and have led to the suggestion that CFS is a disease of chronic immune activation. Some external insult such as an infectious agent, environmental toxin, or stress, sets up conditions in a susceptible individual that lead to a chronic reactivation of latent viruses, which in turn leads to a chronic activation of the immune system, and then cytokines produced by that chronic activation give rise to the symptoms of CFS.

In order to address these questions, we designed a case-control study using CFS patients from our surveillance system. Twenty-six CFS patients were recruited from the Atlanta component of the surveillance system, and 50 controls matched by age, race and sex were selected by random digit dialling. The goals of the study were to (1) determine epidemiologic risk factors for CFS, (2) determine if we could detect a common exposure to an infectious agent and (3) identify differences in immunologic parameters between cases and controls. By studying such a wide range of factors in a small but well-characterized group of cases, we hoped to be able to confirm some of the many smaller studies that exist in the literature.

We found no differences in seroprevalence between cases and controls in any of the agents that we screened. These included enteroviruses, retroviruses, arboviruses, hepatitis B and C, adenovirus, parainfluenza viruses 1,2 and 3, respiratory syncytial virus, parvovirus B19, corona virus, measles, rubella, *Rickettsiaceae, Bartonellae, Borellia burgdorferi, Candida albicans*, and *Chlamydia species*. We also compared both exposure and antibody titers to herpeseviruses, including EBV (early antigen, nuclear antigen and viral capsid antigen), cytomegalovirus (CMV), human herpesvirus 6 (HHV6) and varicella zoster. Again, no differences were seen between cases and controls. Peripheral blood lymphocytes were cultured for retroviruses and HHV-6, and stool samples examined by reverse transcriptase PCR for evidence of enteroviral infection. There was no evidence of infection with these agents in either cases or controls (15). Thus, we conclude that in this group of patients, there was no evidence of a single infectious agent associated with CFS, nor was there evidence of elevated antibody titers consistent with either reactivation of a latent virus or with generalized immune activation.

The immunologic findings from this study are noteworthy for the lack of differences found between cases and controls. We compared proliferation responses to mitogens and antigens, natural killer (NK) cell number and function, cytokine production in response to mitogen, circulating cytokine levels, and an extensive panel of lymphocyte cell surface markers and activation markers. In addition, we determined delayed-type hypersensitivity responses using the Merieux multitest (Connaught Laboratories, Swiftwater, PA), and we examined the level of allergies in the population by screening sera against a panel of allergens specific for the Atlanta area using a radioallergosorbent (RAST) assay. We were unable to confirm reports of increased frequency of allergy and anergy in CFS patients. We were also unable to confirm reports of altered lymphocyte subsets, of decreased natural killer cell function, of decreased lymphocyte proliferative responses, and of altered cytokine profiles. In particular, we found no evidence for circulating cytokines, at least among those we examined, which included IL-1α, IL-1β, IL-2, IL-4, IL-6, interferon-γ, TNF-α, TNF-β and TGF-β (16).

We then subgrouped our patients according to their type of disease onset. Nine CFS patients described a sudden onset to their illness, with fatigue and other symptoms appearing over 1-2 days, while the remaining 17 reported a gradual onset of symptoms. When we repeated our analyses, patients with a sudden illness onset had an increased percentage of CD8 T cells expressing CD11b (suppressor T cell subset). This T cell subset is elevated in acute viral infection. They also showed an increased production of IL-2 in response to mitogen stimulation and an increased proliferative response to *Candida* antigen. In contrast, patients with a gradual disease onset did not show these differences, but showed a decrease in IL-1β production in response to mitogen and had a lower percentage of CD56 (NK) cells expressing CD2 (T cell marker). Three patients in this group also had detectable circulating TGF-β (16). This is not a pattern of alterations usually seen in infectious disease, and may reflect a different pathogenic process. Risk factor analysis also detected differences between cases with gradual illness onset and their matched controls. Gradual onset cases had a greater number of stressful life events in the year prior to illness onset. They were also more likely to have been exposed to agents such as pesticides, to have had a hysterectomy, and to report having symptoms of an upper respiratory tract infection. No risk factors were identified for the sudden onset cases (17).

Taken together, these data suggest that CFS cases can be subgrouped on the basis of the type of disease onset. This in turn may reflect 2 different etiologies for the disease. Sudden onset cases may have an infectious trigger, whereas gradual onset cases may have either an environmental trigger or a stress-induced illness. The numbers in this study are small, and the results need to be confirmed in other studies. However, if there are 2 different processes at work, this would help to explain why there are so many conflicting reports in the literature, since any given group of CFS patients would contain different ratios of the two. This study suggests that subgrouping of CFS patients should be included in future work.

REFERENCES

1. G.P. Holmes, J.E. Kaplan, N.M. Gantz et al. Ann Intern Med, 108, 387-389 (1988).

2. K. Fukuda, S.E. Straus, I. Hickie et al. Ann Intern Med, 121, 953-59 (1994).

3. M. Reyes, H.E. Gary, J.G. Dobbins et al. MMWR CDC Surveill Summ (in press).

4. A.C. Mawle, M. Reyes and D.S. Schmid. Infect Agents and Dis, 2, 333-341 (1994).

5. A.L. Landay, C. Jessop, E.T. Lennette and J.A. Levy. Lancet, 338, 707-712 (1991).

6. S.E. Straus, S. Fritz, J.K. Dale, B. Gould and W. Strober. J Clin Immunol, 13, 30-40, (1993).

7. M. Caligiuri, C. Murray, D. Buchwald, et al. J Immunol, 139, 3306-3313 (1987).

8. N.G. Klimas, F.R. Salvato, R. Morgan and M.A. Fletcher. J Clin Microbiol, 28, 1404-1410 (1992).

9. E. Barker, S.F. Fujimura, M.B. Fadem, A.L. Landay, and J.A. Levy. CID, 418, S136-S141 (1994).

10. E.A. Ojo-Amaize, E.J. Conley and J.B. Peter. CID, 18, S157-S159 (1994).

11. L.J.A. Morrison, W.H.M. Behan and P.O. Behan. Clin exp immunol, 83, 441-446 (1991).

12. U. Tirelli, G. Marotta, S. Improta and A. Pinto. Scand J Immunol, 40, 601-608 (1994).

13. S. Gupta and B. Vayuvegula. Scand J Immunol, 33, 319-327 (1991).

14. A. Lloyd, I. Hickie, C. Hickie, J. Dwyer and D. Wakefield. Clin exp Immunol, 87, 76-79 (1992).

15. A.C. Mawle, R. Nisenbaum, J.G. Dobbins et al. CID 21, 1386-1389 (1995).

16. A.C. Mawle, R. Nisenbaum, J.G. Dobbins et al. JID (in press).

17. M. Reyes, J.G. Dobbins, A.C. Mawle et al. JCFS (in press).

SUPERANTIGENS: THEIR ROLE IN INFECTIOUS DISEASES

Dennis L. Stevens, Ph.D., M.D.
Chief, Infectious Diseases
Veterans Affairs Medical Center
Boise, Idaho 83702

Professor of Medicine
University of Washington School of Medicine
Seattle, Washington

Abstract

In the last 10 years many of the superantigens of the microbial world have been defined and the mechanisms of cellular interaction between lymphocytes and antigen presenting cells has been elucidated in great detail. The consequences of superantigen stimulation of the immune system, though less well defined, can be considered in three separate stages: T-cell proliferation, apoptosis, and recovery. Understanding these stages may explain why diverse superantigens may cause markedly different clinical processes ranging from acute shock to chronic arthritis and may form the basis for novel treatments of these diverse diseases.

Introduction

Until very recently, the T-lymphocyte was not considered a significant contributor to the pathogenesis of bacterial infections. Currently, there is widespread interest in defining the role of the lymphocyte in infections caused by Gram positive pathogens such as Staphylococcus aureus and Streptococcus pyogenes. Most of these investigations have been done in vitro and there is now firm understanding of the molecular mechanisms of activation of T-lymphocytes by proteins produced by such bacteria. Indeed the list of potential superantigens gets longer daily as many investigators attempt to demonstrate that their favorite organism produces proteins that can interact with T-lymphocytes in this unique way. Thus, a barrage of papers have somewhat confused the literature. Indeed it is difficult to imagine that the final common immunological pathway is similar for diverse diseases such as staphylococcal food poisoning, staphylococcal toxic shock syndrome, streptococcal toxic shock syndrome, arthritis caused by Mycobacterium species and mouse mammary tumor virus infection. Though a common mechanism may not explain all these diseases, is there evidence that superantigens play significant role in the pathogenesis of any disease? This manuscript will review the known characteristics of superantigens and will explore the cellular and molecular mechanisms that affect the T-lymphocyte.

Superantigens: Products of microbes.

The superantigens studied to date include exotoxins from Staphylococcus aureus, and Streptococcus pyogenes, some surface components of S. pyogenes, Mycoplasma, and Yersinia, and some viruses. Staphylococcal enterotoxin B (SEB) is without question the most studied bacterial superantigen. The streptococcal pyrogenic exotoxins (SPE) A, B and C as well as two newly described streptococcal superantigens, streptococcal superantigen (SSA) and mitogenic factor (MF), also have been shown to have properties of superantigens. In addition, there is some evidence that certain streptococcal M-proteins (such as M-5 protein) may also serve as superantigens. Mycoplasma arthriditis produces surface proteins that have superantigen qualities which may play a significant role in experimental arthritis due to this microbe. One other surface component, cytoplasmic factor of Yersinia enterocolitica, has been shown to have superantigenic properties. Finally two viruses, mouse mammary tumor virus and the human immunodeficiency virus, are putative superantigens.

The Consequences of Superantigen Interaction with the Immune System.

Cells of the immune system are constantly bombarded with antigens. Though most are recognized as foreign, clearly self antigens interface with immune cells as well. The dynamic impact of superantigen stimulation of the immune system can be considered in three separate parts: early events, subacute events, and late events. The early events require that the putative superantigen interact simultaneously with both the MHC Class II complex of an antigen presenting cell and precise areas of the T-cell receptor called Vβ regions. There is great interest in defining the structure/function relationships of different superantigens and in elucidating the precise binding and activational regions of the MHC Class II and T-cell receptor, however the focus of this manuscript will be to discuss the consequence of superantigen-T lymphocyte interaction. Following recognition, proliferation of lymphocytes bearing specific Vβ regions of the T-cell receptor (lymphocyte blastogenesis) occurs with early production of monocyte-derived cytokines followed later by production of cytokines of lymphocyte origin.

In the sub-acute stage, the expanded population of T-lymphocytes becomes tolerant to subsequent stimulation by homologous and heterologous superantigen. In addition, a general state of anergy exists. This phenomenon occurs both in vitro and in vivo. During this time, there is also evidence that the expanded clone of T-lymphocytes undergoes a diminution in size, referred to as deletion. This may occur through a process of programmed cell death termed apoptosis. Finally, this stage is associated with down regulation of both IL-2 production and IL-2 receptor (IL-2R) expression.

The late stage of superantigen activation is characterized by the return of normal lymphocyte function. Thus, the number of lymphocytes in general as well as the number of T-lymphocyte subsets bearing specific Vβ repertoires returns to normal. In addition, specific anergy and tolerance are resolved. Finally, antibody production, stimulated by classical antigen processing proceeds in a normal fashion.

Recent studies have shown that activation of the T-cell via the superantigen mechanism requires the appropriate toxin, viable T lymphocytes bearing a specific Vβ repertoire, and either a viable antigen presenting cell (APC), a non-viable APC, or a fragment of the MHC Class II receptor (1). Other accessory molecules may also participate in the docking between MHC Class II and T-cell receptor (TCR). One of the T-lymphocyte binding sites for enterotoxin appears to be CD4 (2). This could serve to stabilize the complex between the TCR and MHC Class II. Further, blocking of other surface molecules such as ICAM-1, CD11a, CD28, or CD2 with monoclonal antibodies effectively prevented T-cell proliferation in the presence of staphylococcal superantigens (3).

The consequences of superantigen stimulation of the immune system are T-lymphocyte proliferation and generation of vast quantities of the lymphokines IL-2, gamma interferon and TNFβ (4-6) as well as the monokines, TNFα, IL-1 and IL-6 (4,7), yet the dynamics of production are quite different for each. Specifically, during the first 24 hours, production of TNFα by peripheral blood mononuclear cells (PBMC) predominates. By 48 hours, TNFβ is detectable, and at 72 hours, equal quantities of TNFα and TNFβ can be measured (4). IL-2, a lymphocyte growth factor, is believed to drive the proliferation of lymphocytes.

The control mechanisms of cytokine production via the superantigen mechanism is poorly understood, however both counter-regulatory cytokines, shedding or internalization of receptors, and accelerated programmed cell death are likely involved. The conventional wisdom regarding T-cell proliferation states that the initial proliferation of T-cells in response to superantigens is followed by the rapid depletion of T-cell subsets bearing the specific Vβ repertoire through accelerated programmed cell death or apoptosis (5).

The Immunopathogenesis of Staphylococcal Toxic Shock Syndrome

Recently, T-lymphocyte dynamics were studied in a mouse model of staphylococcal enterotoxin B (SEB)-induced shock (8,9). There was clear evidence of Vβ-selective clonal expansion 24 - 48 hours after administration of SEB (8,9). On day 3, clonally expanded cells became depleted through accelerated programmed cell death (apoptosis) resulting in anergy (8,9). McCormick et al (10) demonstrated that this expansion-deletion sequence occurs predictably in vivo when high concentrations of a superantigen are used. However, deletions also occur using low dosages. In fact, concentrations which were insufficient to cause T-cell blastogenesis in vivo were still capable of depleting specific Vβ subsets of T-lymphocytes (10). Using sub-lethal concentrations of a superantigen, Miethke recently demonstrated that 4 - 6 hours after in vivo challenge with SEB, BALB/c mice became resistant to an otherwise lethal dose of SEB plus D-galactosamine (11). These authors demonstrated that PBMC harvested from such animals during this time lost the ability to release lymphokines such as TNF and IL-2 (11). They hypothesized that endogenous corticosteroids released in response to the first wave of lymphokines induced by SEB may have suppressed T-cell responsiveness. Though this might be an adequate explanation, counter-regulatory cytokines likely play a more important role. For example, 10 ng/ml of IL-10, a regulatory cytokine produced by the Th2 subset of CD4 + helper T-lymphocytes, inhibited synthesis of TNFα and gamma interferon production by TSST-1-stimulated PBMC by 68 and 86% respectively (12). The relevance of this endogenous control mechanism in vivo is uncertain, however, since stimulation of PBMC with TSST-1 induces only 72 pg/ml of IL-10 (12), a concentration well below that required to inhibit cytokine production. Whether or not IL-10 contributes as an autocrine or paracrine endogenous inhibitor of PBMC synthesis of TNFα or TNFβ, Krauakauer et al point out that exogenous administration of recombinant IL-10 might be a rational treatment for superantigen-induced disease (12). Recent studies showing that IL-10 protected animals from lethal challenge supports this hypothesis (13). Interestingly, IL-10 did not inhibit T-lymphocyte blastogenesis (12), suggesting that IL-10 works largely by inhibiting transcriptional events.

Though the previously described animal models of superantigen injection demonstrate lethality, what is the evidence that S. aureus toxins such as TSST-1 or the enterotoxins cause an illness resembling Staph TSS? The animal model that has been best studied in this regard is the rabbit whiffle ball model. Briefly, sterile plastic chambers are implanted beneath the skin of rabbits and allowed to mature over the course of 3 - 4 weeks. Strains of S. aureus are then inoculated into these chambers and physiologic measurements of blood chemistries, blood pressure, pulse, temperature and mortality are recorded. Scott and co-workers (14) and Rasheed et al (15) using this model, and Azevedo et al (16) using an implanted uterine diffusion chamber model, demonstrated that strains producing TSST-1 were more likely to induce a toxic shock-like syndrome than strains which did not produce TSST-1. Recently, it was shown that strains of S. aureus which produced altered TSST-1 toxins lacking the ability to induce mitogenicity were non-toxic in the rabbit whiffle ball chamber model (17), though the reason for the loss of mitogenicity was unexplained. Subsequently, using both genetically altered SEC mutants and peptide fragments of SEC to stimulate lymphocyte proliferation in vitro, it was determined that toxin binding to the alpha helix of the MHC Class II portion of the antigen presenting cell was critically important in the induction of lymphocyte blastogenesis (18).

The role of TSST-1 in inducing a TSS-like illness is also supported by the demonstration that a neutralizing monoclonal antibody prevented illness in rabbits challenged with a TSST-1 producing strain of S. aureus (19). Parsonnet et al (20) has also developed a slow infusion model which allows the investigation of purified, recombinant, and genetically altered toxins.

The Immunopathogenesis of Streptococcal Toxic Shock Syndrome (Strep TSS).

It has been suggested that Strep TSS could, like Staph TSS, be related to the ability of certain streptococcal factors to act as "superantigens" (21). There is data to suggest that SPEA, SPEB, SPEC, MF, SSA and M-protein can stimulate T cell responses through their ability to bind to both the Class II MHC complex of antigen presenting cells and the Vβ region of the T cell receptor (21) and production of cytokines capable of mediating shock and tissue injury. Recently, Hackett and Stevens demonstrated that SPEA induced both TNFα and TNFβ from mixed cultures of monocytes and lymphocytes (22), supporting the role of these lymphocyte derived cytokines in shock associated with strains producing SPEA. Dale et al (23) demonstrated that purified polypeptide fragments of pepsin extracted M proteins induced blastogenic responses in human lymphocytes but not laboratory animals. Kotb subsequently (24) has shown that a digest of M-protein type 5 (pep M5) can also stimulate T cell responses by this superantigen mechanism. Induction of cytokines by monocytes was dependent upon the presence of T-lymphocytes (24). Interestingly, the requirement for intact T-cells could be overcome by the addition of exogenous interferon gamma (24). Pep M5 also stimulated mixed cultures of monocytes and lymphocytes to produce the lymphokines, interferon gamma and TNFβ, though TNFβ production was greatly enhanced by the addition of exogenous gamma interferon (24). These in vitro results suggest that superantigens could play a role in the pathogenesis of Strep TSS. Proof would require demonstration of massive expansion of T-cell subsets bearing Vβ repertoires specific for the putative superantigen(s). Recently, quantitation of such T-cell subsets in patients with acute Strep TSS demonstrated deletion rather than expansion, suggesting that perhaps the life-span of the expanded subset was shortened by a process of apoptosis (25). In addition, the subsets deleted were not specific for SPEA, SPEB, SPEC, or MF suggesting that perhaps an as yet undefined superantigen may play a role (25)].

Cytokine production by less exotic mechanisms may also contribute to the genesis of shock and organ failure in Staph TSS and Strep TSS. Peptidoglycan, lipoteichoic acid (26) and killed streptococci (27,28) are capable of inducing TNFα production by mononuclear cells in vitro (4,28,29). Exotoxins such as streptolysin O (SLO) are also potent inducers of TNFα and IL-1β. SPEB, a proteinase precursor, has the ability to cleave pre-IL-1β to release preformed IL-1β (30). Finally, SLO and SPEA together have additive effects in the induction of IL-1β by human mononuclear cells (22). Whatever the mechanisms, induction of cytokines in vivo is likely the cause of shock, and many fluid phase streptococcal virulence factors as well as cell wall components are potent inducers of TNF and IL-1. Thus, the specific virulence factors that induce cytokine production in patients may be complex, particularly among the 60% of Strep TSS patients who are bacteremic (31). Here the systemic immune system would be exposed to a barrage of streptococcal virulence factors all of which are capable of cytokine induction through a variety of cellular mechanisms.

Summary

The interaction between microbial virulence factors and an immune or non-immune host determines the epidemiology, clinical syndrome and outcome of infection. Since horizontal transmission of group A streptococcus is well documented, the only explanation for the absence of a high attack rate of invasive infection is the presence of significant herd immunity against one or more of the virulence factors responsible for Strep TSS (32). This model explains why (1) epidemics have not occurred and (2) why a particular strain of group A streptococcus can cause different clinical manifestations in the same community (33). Similarly, the presence or absence of neutralizing antibody against TSST-1 may determine which individuals may develop Staph TSS. Alternatively, there may be certain individuals in communities that have a predisposition to a violent reaction to staphylococcal or streptococcal virulence factors such as superantigens. It would be expected that these individuals might have populations of T-lymphocytes that harbor particular Vβ regions of the T-cell receptor. Finally, though the superantigen mechanism is an attractive explanation for the fulminant onset of shock and organ failure, demonstration of this process in humans awaits a definitive study.

References

1. R.H. See, W.W.S. Kum, A.H. Chang, S.H. Goh and A.W. Chow. Infect Immun, 60, 2612-2618 (1992).
2. S. Bavari and R.G. Ulrich. Infect Immun, 63, 423-429 (1995).
3. T. Krakauer. J Leuk Biol, 56, 458-463 (1994).
4. S.P. Hackett and D.L. Stevens. J Infect Dis, 168(1), 232-235 (1993).
5. P. Marrack and J.W. Kappler. Science, 248, 705-711 (1990).
6. A. Norrby-Teglund, M. Norgren, S.E. Holm, U. Andersson and J. Andersson. Infect Immun, 62, 3731-3738 (1994).
7. D.J. Fast, P.M. Schlievert and R.D. Nelson. Infect Immun, 57, 291-294 (1989).
8. H.R. MacDonald, S. Baschieri and R.K. Lees. Eur J Immunol, 21, 1963-(1991).
9. Y. Kawabe and A. Ochi. Nature, 349, 245-(1991).
10. J.E. McCormack, J.E. Callahan, J. Kappler and P.C. Marrack. J Immunol, 150, 3785-3792 (1993).
11. T. Miethke, C. Wahl, K. Heeg and H. Wagner. J Immunol, 150, 3776-3784 (1993).
12. T. Krakauer. J Infect Dis, 172, 988-992 (1995).
13. A.G.D. Bean, R.A. Freiberg, S. Andrade, S. Menon and A. Zlotnik. Infect Immun, 61, 4937-4939 (1993).
14. D.F. Scott, J.M. Kling, J.J. Kirkland and G.K. Best. Infect Immun, 39, 383-387 (1983).
15. J.K. Rasheed, R.J. Arko, J.C. Feeley, F.W. Chandler, C. Thornsberry, R.J. Gibson, M.I. Cohen, C.D. Jeffries and C.V. Broome. Infect Immun, 47, 598-604 (1985).
16. J.C.S. De Azavedo, T.J. Foster, P.J. Hartigan, J.P. Arbuthnott, M. O'Reilly, B.N. Kreiswirth and R.P. Novick. Infect Immun, 50, 304-309 (1985).
17. P.F. Bonventre, H. Heeg, C. Cullen and C. Lian. Infect Immun, 61, 793-799 (1993).
18. M.L. Hoffmann, L.M. Jablonski, K.K. Crum, S.P. Hackett, Y. Chi, C.V. Stauffacher, D.L. Stevens and G.A. Bohach. Infect Immun, 62, 3396-3407 (1994).
19. G.K. Best, D.F. Scott, J.M. Kling, M.R. Thompson, L.E. Adinolfi and P.F. Bonventre. Infect Immun, 56, 998-999 (1988).
20. J. Parsonnet, Z.A. Gillis, A.G. Richter and G.B. Pier. Infect Immun, 55, 1070-1076 (1987).
21. J.A. Mollick and R.R. Rich. Clin Res, 39(2), 213A-(1991).
22. S.P. Hackett and D.L. Stevens. J Infect Dis, 165, 879-885 (1992).
23. J.B. Dale, W. Simpson, I. Ofek and E.H. Beachey. J Immunol, 126, 1499-1505 (1981).
24. M. Kotb, M. Tomai, G. Majumdar, J. Walker and E.H. Beachey. XI Lancefield International Symposium on Streptococcal Diseases, Sienna, Italy(1990).
25. R. Watanabe-Ohnishi, D.E. Low, A. McGeer, D.L. Stevens, P.M. Schlievert, D. Newton, B. Schwartz, B. Kreiswirth, Ontario Streptococcal Study Project and M. Kotb. J Infect Dis, 171, 74-84 (1995).
26. D.L. Stevens, A.E. Bryant and S.P. Hackett. Curr Opin Infect Dis, 5(3), 355-363 (1992).
27. S. Hackett, J. Ferretti and D. Stevens. Program and Abstracts of the American Society for Microbiology, Abstract B-249(1994).
28. H. Muller-Alouf, J.E. Alouf, D. Gerlach, J.H. Ozegowski, C. Fitting and J.M. Cavaillon. Infect Immun, 62, 4915-4921 (1994).
29. D.L. Stevens. Clin Infect Dis, 14, 2-13 (1992).
30. V. Kappur, M.W. Majesky, L.L. Li, R.A. Black and J.M. Musser. Proc Natl Acad Sci USA, 90, 7676-7680 (1993).
31. D.L. Stevens, M.H. Tanner, J. Winship, R. Swarts, K.M. Reis, P.M. Schlievert and E. Kaplan. N Eng J Med, 321(1), 1-7 (1989).
32. S.E. Holm, A. Norrby, A.M. Bergholm and M. Norgren. J Infect Dis, 166, 31-37 (1992).
33. D.L. Stevens. Pediatr Infect Dis J, 13, 561-566 (1994).

SEARCHING FOR SUPERANTIGENS

P.M. Schlievert
Department of Microbiology
Medical School, University of Minnesota
Minneapolis, Minnesota 55455

ABSTRACT

Superantigens comprise a large group of viral and bacterial proteins that stimulate T lymphocyte proliferation without regard for the antigenic specificity of the T cells but dependent on the composition of the variable part of the β chain of the T cell receptor. Superantigens induce T cell proliferation dependent on class II MHC molecules on antigen presenting cells but do not require processing. Major subfamilies of superantigens include the viral superantigens, the bacterial pyrogenic toxin superantigens, and other bacterial superantigens. Two major approaches have been taken to identify superantigen association with human diseases: a) assessing Vβ T cell receptor skewing in peripheral blood or tissues of patients with illnesses, b) recognition of toxic shock syndrome and related illnesses which are likely to be caused by superantigens.

INTRODUCTION

Superantigens comprise a large family of microbial products that stimulate massive T lymphocyte proliferation in an unusual way (1). A list of T cell superantigens is presented in TABLE I.

The large group of superantigens can be divided into subfamilies that include viral and bacterial superantigens. The bacterial superantigens may be again further subdivided into pyrogenic toxin superantigens and those that are not pyrogenic toxins (2-8).

Pyrogenic toxin superantigens (PTSAgs) are the most well known group of superantigens, including staphylococcal toxic shock syndrome toxin-1 (TSST-1) and enterotoxins and group A streptococcal pyrogenic exotoxins (SPEs, scarlet fever toxins). The PTSAgs are distinguished from other superantigens in that PTSAgs have numerous biological properties (2) in addition to superantigenicity (TABLE II). There is no evidence other superantigens have these other

TABLE I

Microbial Superantigens

I. Viral

A. Murine mammary tumor virus

B. Rabies virus

C. Epstein Barr virus

D. HIV

II. Bacterial

A. Pyrogenic toxin superantigens
1. Staphylococcal toxic shock syndrome toxin-1
2. Staphylococcal enterotoxins serotypes A-H but not F
3. Group A streptococcal pyrogenic exotoxins serotypes A-C
4. Group A streptococcal superantigen
5. Group A streptococcal pyrogenic exotoxin F (mitogenic factor)

B. Staphylococcal exfoliative toxins A and B

C. *Mycoplasma arthritidis* mitogen

D. *Yersinia enterocolitica* superantigen

E. *Yersinia pseudotuberculosis* mitogen

TABLE II

Biological activities of pyrogenic toxin superantigens

1. Superantigenicity - cytokine release
2. Pyrogenicity
3. Enhancement of:
    a. endotoxin shock
    b. type I hypersensitivity
    c. cardiotoxicity
4. Lethality in miniosmotic* pumps
5. Ability to bind endothelial cells
6. Reactivation of arthritis in rats
7. Ability to bind to endotoxin
8. Host cell lethality in the presence of endotoxin
9. Emesis - enterotoxins only

* Miniosmotic pumps, containing pyrogenic toxin superantigens, are implanted subcutaneously in rabbits. The pumps release constant amounts of toxin over a 7 day time period, thus continuously exposing the animals to toxin. Rabbits typically succumb to 100μg of toxin administered by this route in two major groups, one on day 1 to 2, and a second on day 4 to 5.

properties. Furthermore, PTSAgs have all been associated with causation of life threatening toxic shock syndromes (TSS), whereas other superantigens have not (2).

T cell superantigens stimulate the immune system differently than typical antigens, which stimulate only approximately 1/10,000 T cells, and true nonspecific mitogens such as concanavalin A which stimulate all T cells to proliferate. The defining properties of superantigens are shown in TABLE III.

Superantigens require interaction with the class II MHC molecules on antigen presenting cells (APCs) for maximal T cell proliferation, but processing by APCs is not necessary (1). Superantigens appear to be able to interact with nearly any class II MHC molecule, but some interaction preference is seen, in particular, interaction with HLA DR molecules is preferred.

Superantigens interact with the T cell receptor outside of the typical antigenic peptide groove and dependent on the composition of the β chain of the variable region of the T cell receptor complex (Vβ TCR). Thus, for example, TSST-1 stimulates Vβ2 containing T cells to proliferate in TSS patients such that 60-70% of the patients' peripheral blood T cells are Vβ2 (9).

SEARCHING FOR SUPERANTIGENS

Two major approaches have been taken to search for superantigen causation of human diseases. These are discussed below.

First, the dramatic Vβ T cell skewing induced by superantigens has been used in attempt to identify causative factors in numerous human diseases. A partial listing of human diseases that are/may be induced or reactivated by superantigens is shown in TABLE IV.

Examples of the use of this approach are the association of Kawasaki syndrome and guttate psoriasis with superantigens. Recent studies have shown that patients with Kawasaki syndrome have significant Vβ2 TCR and to a lesser extent Vβ8 TCR skewing in peripheral blood and tissues, suggesting superantigen involvement (10). Subsequent studies have suggested bacterial toxins such as TSST-1, SPE type C, and exfoliative toxins, all of which cause characteristic signature Vβ2 TCR skewing, may contribute to the illness (11). It is important to note that these studies are controversial in that other studies have failed to show either Vβ TCR skewing in Kawasaki syndrome or association with bacterial toxins (12, 13). This remains an active area of research.

Two studies have shown significant Vβ TCR skewing in association with guttate psoriasis (14, 15). These patients typically are infected with group A streptococci one to two weeks prior to onset of symptoms. Examination of group A streptococci from guttate psoriasis patients demonstrates their production of superantigens capable of causing the Vβ TCR skewing seen, thus leading to the hypothesis the toxins contribute to the illness.

TABLE III

Mechanisms of T cell stimulation by superantigens
1. Superantigens do not require processing by antigen presenting cells.
2. For optimal T cell stimulation, superantigens interact with both class II MHC molecules on presenting cells and the T cell receptor.
3. Superantigens stimulate T cell proliferation independent of antigen specificity but dependent on the composition of the variable region of the β chain of the T cell receptor.

TABLE IV

Known and possible disease association with superantigens
I. Acute diseases
A. Staphylococcal toxic shock syndrome
B. Streptococcal toxic shock syndrome
C. Scarlet fever
D. Staphylococcal scalded skin syndrome
E. Kawasaki syndrome
F. Fevers of unknown origin
II. Chronic diseases
A. Arthritis
B. Atopic dermatitis
C. Guttate psoriasis
D. Rheumatic fever
E. Multiple sclerosis

The use of Vβ TCR skewing to identify potential causative superantigens may be difficult because of variability in degree of skewing seen. It is clearly established that staphylococcal TSS is caused by TSST-1 and enterotoxins. It is also likely that superantigen contribute significantly to streptococcal TSS. It is important to note, however, that despite the massive T cell proliferation that may be induced by causative superantigens, the time frame of examining patients for Vβ TCR skewing is very important. Dependent on the time evaluated, significant Vβ TCR amplification, normal numbers of T cell subpopulations, or Vβ TCR depletion may be seen. As an example, in a recent article, significant Vβ TCR depletion was seen in patients with serious streptococcal disease as opposed to T cell proliferation and Vβ TCR skewing (16).

Possibly the most definitive method for association of superantigens with illnesses is to evaluate TSS and scarlet fever-like illnesses for causative toxins. The toxins identified to date belong to the large family of PTSAgs and thus are characterized by their shared physicochemical and biological properties.

The following methods are being used in our laboratory in attempt to identify PTSAgs that may cause TSS illnesses in patients who are not infected with either *S. aureus* or group A streptococci. We have received more than 30 group B, 30 group G, and 10 group C streptococcal isolates from TSS patients (groups B isolates were not from neonatal sepsis patients). In addition, two each of group F streptococcus, enterococci, and *Streptococcus pneumoniae* and a large number of α hemolytic streptococci and coagulase negative staphylococci have been submitted. Finally, it is well known the *Bacillus cereus* strains have the ability to cause food poisoning nearly indistinguishable from staphylococcal food poisoning and may make PTSAgs related to staphylococcal enterotoxins. TABLE V provides a listing of bacteria, other than *Staphylococcus aureus* and group A streptococci, that are most promising sources of superantigens.

PTSAgs are all secreted proteins with molecular weights of 20-30,000 and are resistant to denaturation by ethanol (12). Partial purification of the toxins from culture fluids is easily achieved by precipitation with 75% (final concentration) ethanol. Resolubilized material is then tested for ability to stimulate lymphocyte proliferation (17), induce fever (18), and enhance susceptibility to lethal endotoxin shock (19).

Positive culture material from the above quick screens are then subjected to preparative thin layer isoelectric focusing and fractions tested for lymphocyte proliferative activity. In our laboratory, we use rabbit splenocytes as the cells for screening fractions, since rabbits are useful models for the study of TSS illnesses. By these methods, we have shown that group B, C, F, G and some α hemolytic streptococci make possible superantigens that may contribute to patients' illnesses.

TABLE V

Other bacteria that are likely to make superantigens
A. Coagulase negative staphylococci
B. β hemolytic streptococci
1. Group B
2. Group C
3. Group F
4. Group G
C. α hemolytic streptococci (including *S. pneumoniae*)
D. Enterococci
E. *Bacillus cereus*

It is interesting to note that the majority of bacterial superantigens are obtained from Gram positive bacteria. It is possible that the Gram positive association resides in the 100,000 to million fold enhancement of endotoxin shock associated with PTSAgs. It is likely that production of PTSAgs by Gram negative organisms, which contain endotoxin, would likely yield such highly lethal pathogens, the organisms would be evolutionarily selected against.

By the rapid screening methods outlined above, suggestive evidence of superantigen production by organisms is obtained. However, final proof of superantigen production depends on: a) establishing the proteins cause proliferation of T cells dependent on Vβ TCR, and b) showing the T cell proliferation induced is not antigen specific.

Two major methods have been used to establish Vβ TCR dependent proliferation. This is accomplished by either use of flow cytometry after staining stimulated T cells with antibodies specific for individual Vβs (19), or by a semiquantitative PCR specific for the Vβs (16). Both methods have been used to reliably demonstrate Vβ TCR skewing.

Once the Vβ TCR dependent proliferation is established, it is important to show the proliferating T cells are polyclonal rather than high level stimulation of a single antigen specific T cell clone. This is accomplished by two methods: a) sequencing the V, D, J junction fragments of the β chain of the TCR to show variability and thus polyclonality, b) stimulation of T cell clones that are known to be specific for other antigens (20).

As a final comment: after characterization of superantigens made by organisms, their role in diseases still remains to be established. With the exception of certain PTSAgs and exfoliative toxins associated with TSS-like illnesses and scalded skin syndrome, respectively, definitive establishment of superantigen causation of human illnesses has been difficult. Major studies continue in these areas.

REFERENCES

1. P. Marrack and J. Kappler. Science 248: 705-711 (1990).

2. G.A. Bohach, D.J. Fast and R.D. Nelson. Crit Rev Microbiol 17: 251-272 (1990).

3. D.A. Blomster-Hautamaa and P.M. Schlievert, in Handbook of natural toxins, Vol. 4, M.C. Hardegree and A.T. Tu, eds, Marcel Dekker, Inc., New York, (1988) pp. 297-330.

4. B.C. Cole, D.R. Kartchner and D.J. Wells. J Immunol 142: 4131-4127 (1989).

5. P.M. Stuart and J.G. Woodward. J Immunol 148: 225-233 (1992).

6. J. Abe, T. Takeda, Y. Watanabe, H. Nakao, N. Kobayashi, D.Y.M. Leung and T. Kohsaka. J Immunol 151: 4183-4188 (1993).

7. J.A. Mollick, G.G. Miller, J.M. Musser, R.G. Cook, D. Grossman and R.R. Rich. J Clin Invest 92: 710-719 (1993).

8. A. Norrby-Teglund, D. Newton, M. Kotb, S.E. Holm and M. Norgren. Infect Immun 62: 5227-5233 (1994).

9. Y. Choi, J.A. Lafferty, J.R. Clements, J.K. Todd, E.W. Gelfand, J. Kappler, P. Marrack and B.L. Kotzin. J Exp Med 172: 981-984 (1990).

10. J. Abe, B. L. Kotzin, K. Jujo, M.E. Melish, M.P. Lode, T. Kohasaka, D.Y.M. Leung. Proc Natl Acad Sci 89: 4066-4070 (1992).

11. D.Y.M. Leung, H.C. Meissner, D.R. Fulton, D.L. Murray, B.L. Kotzin and P.M. Schlievert. Lancet 342: 1385-1388 (1993).

12. B.A. Pietra, J. De Inocenio, E.H. Giannini and R. Hirsch. J Immunol 153: 1881-1888 (1994).

13. A. Nishiyori, M. Sakaguchi, H. Kato, H. Igarashi and K. Miwa. Lancet 343: 300 (1994).

14. H.M. Lewis, B.S. Baker, S. Bokth, A.V. Powles, J.J. Garioch, H. Valdimarsson and L. Fry. Br J Dermatol 129: 514-520 (1993).

15. D.Y.M. Leung, J.B. Travers, R. Giorno, D.A. Norris, R. Skinner, J. Aelion, L.V. Kazemi, M.H. Kim, A.E. Trumble, M. Kotb and P.M. Schlievert. J Clin Invest 96: 2106-2112 (1995).

16. R. Watanabe-Ohnishi, D.E. Low, A. McGreer, D.L. Stevens, P.M. Schlievert, D. Newton, B. Schwartz, B. Kreiswirth, Ontario Streptococcal Study Project and M. Kotb. J Infect Dis 171: 74-84 (1995).

17. E.L. Barsumian, P.M. Schlievert and D.W. Watson. Infect Immun 22: 681-688 (1978).

18. P.M. Schlievert, K.N. Shands, B.B. Dan, G.P. Schmid and R.D. Nishimura. J Infect Dis 143: 509-516 (1981).

19. B.A.B. Leonard, P.K. Lee, M.K. Jenkins and P.M. Schlievert. Infect Immun 59: 1210-1214 (1991).

20. S.D. Norton, P.M. Schlievert, R.P. Novick and M.K. Jenkins. J Immunol 144: 2089-2095 (1990).

AUTHOR INDEX

SUBJECT INDEX

About the Editors

DANIEL AMSTERDAM is Professor of Microbiology and Pathology, Associate Professor of Medicine, and a member of the Ernest Witebsky Center for Immunology at the State University of New York at Buffalo School of Medicine and Biomedical Sciences, and Director of Laboratory Medicine and Clinical Microbiology and Immunology at the Erie County Medical Center, Buffalo, New York. Dr. Amsterdam is the author, coauthor, or editor of over 120 scientific papers, book chapters, and review articles and a member of the editorial committees of *Immunological Investigations* (Marcel Dekker, Inc.), *Antimicrobics and Infectious Diseases*, the *Journal of Clinical Microbiology*, and the *Manual of Clinical Microbiology*. He is a Fellow of the American Association for the Advancement of Science, the American Society for Microbiology, and the Infectious Diseases Society of America, and a member of the American Society of Investigative Pathology and the Society for In Vitro Biology, among other organizations. Dr. Amsterdam received the Ph.D. degree (1965) in cellular cytophysiology from New York University, New York, New York.

ROGER K. CUNNINGHAM is Director of the Ernest Witebsky Center for Immunology and Associate Professor of Microbiology at the State University of New York at Buffalo School of Medicine and Biomedical Sciences. He is the author or coauthor of over 40 scientific papers and book chapters; the author, coauthor, or editor of several books; and he serves on the editorial board of *Immunological Investigations* (Marcel Dekker, Inc.). Dr. Cunningham received the Ph.D. degree (1970) in microbiology from the State University of New York at Buffalo School of Medicine.

CAREL J. VAN OSS is Professor of Microbiology; Adjunct Professor of Chemical Engineering and Geology; and a member of the Ernest Witebsky Center for Immunology at the State University of New York at Buffalo. Dr. van Oss is the author or coauthor of over 300 scientific papers and the author, coauthor, or editor of 10 books, including *Interfacial Forces in Aqueous Media*, *Immunochemistry*, and *Transfusion Immunology and Medicine*, Editor-in-Chief of the journals *Immunological Investigations* and *Preparative Biochemistry & Biotechnology*, and Associate Editor of *Separation and Purification Methods* (above titles, Marcel Dekker, Inc.), and a member of the editorial committees of *Colloids and Surfaces—B*, the *Journal of Protein Chemistry*, and *Cell Biophysics*. He is a member of the American Association for the Advancement of Science, the American Association of Immunologists, and the American Chemical Society, among other organizations. Dr. van Oss received the Ph.D. degree (1955) in physical biochemistry from the University of Paris (Sorbonne).